Cyberspace and Outer Space: Transitional Challenges for Multilateral Verification in the 21st Century

Proceedings of the Fourteenth Annual Ottawa NACD Verification Symposium

Edited by: J. Marshall Beier and Steven Mataija

CENTRE FOR INTERNATIONAL
AND SECURITY STUDIES
York University
Toronto, Canada
1997

THE YORK CENTRE FOR INTERNATIONAL AND SECURITY STUDIES supports and encourages innovative theoretical and empirical research in international relations and security studies. The Centre fosters scholarly research of interest to academic and policy communities in Canada and abroad. The Centre also coordinates and facilitates a variety of scholarly activities including conferences, workshops, lectures, seminars, and research publications.

Research at the Centre falls within four streams: (1) Strategic Studies; (2) Political Economy of Security; (3) Regional Conflict; and (4) Non-Traditional Analyses of Conflict and Security.

For further information please contact:

Centre for International and Security Studies
3rd Floor York Lanes
York University
4700 Keele Street
North York, Ontario
Canada M3J 1P3

Telephone:	(416) 736-5156
Fax:	(416) 736-5752
E-mail:	yciss@yorku.ca
Internet:	http://www.yorku.ca/research/ciss

Canadian Cataloguing in Publication Data

Ottawa NACD Verification Symposium (14th : 1997 :
Montebello, Quebec)
Cyberspace and outer space : transitional challenges
for multilateral verification in the 21st century

Proceedings of a symposium held in Montebello, Quebec,
Mar. 12–15, 1997.
Includes bibliographical references.
ISBN 0-920231-24-1

1. Arms control – Congresses. 2. Arms control –
Verification – Congresses. 3. Astronautics, Military –
Congresses. I. Beier, J. Marshall, 1966– .
II. Mataija, Steven, 1968– . III. York Centre for
International and Security Studies. IV. Title.

UA12.5.O77 1997 327.1′74 C97-931154-3

Table of Contents

List of Contributors

PHILLIP J. BAINES, Verification Research Officer, Non-Proliferation, Arms Control and Disarmament Division, Department of Foreign Affairs and International Trade, Ottawa, Canada

JIM BAYER, Program Coordinator, Conflict Analysis and Management, Royal Roads University, Victoria, Canada

J. MARSHALL BEIER, Department of Political Science, Centre for International and Security Studies, York University, Toronto, Canada

CHRISTOPHER BURTON, Researcher, Center for Information Strategy and Policy, Science Applications International Corporation (SAIC), McLean, United States

JASON CAMERON, Visiting Information Officer, The Uranium Institute, London, United Kingdom

F.R. (RON) CLEMINSON, Senior Advisor on Verification, Non-Proliferation, Arms Control and Disarmament Division, Department of Foreign Affairs and International Trade, Ottawa, Canada

JEFFREY R. COOPER, Director, Center for Information Strategy and Policy, Science Applications International Corporation (SAIC), McLean, United States

ALAN CRAWFORD, Research Coordinator (Verification), Non-Proliferation, Arms Control and Disarmament Division, Department of Foreign Affairs and International Trade, Ottawa, Canada

JONATHAN DEAN, Adviser for International Security Issues, Union of Concerned Scientists, Washington, D.C., United States

DAVID DECLERQ, NACD Consultant, Kars, Canada

JAMES DER DERIAN, Department of Political Science, University of Massachusetts, Amherst, United States

THOMAS GRAHAM, JR, Special Representative of the President for Arms Control, Non-Proliferation, and Disarmament, United States Arms Control and Disarmament Agency (ACDA), Washington, D.C., United States

RICHARD GUTHRIE, Verification Technology Information Centre (VERTIC), London, United Kingdom

GEORGE LINDSEY, Senior Research Fellow, Canadian Institute of Strategic Studies, Ottawa, Canada

Steven Mataija, Conference and Information Coordinator, Centre for International and Security Studies, York University, Toronto, Canada

Patricia Bliss McFate, Senior Scientist and Program Director, Center for Global Security and Cooperation, Sante Fe, United States

David Mussington, Senior Policy Analyst, Critical Technologies Institute, RAND Corporation, Washington, D.C., United States

Sergei Plekhanov, Department of Political Science, Centre for International and Security Studies, York University, Toronto, Canada

Brad Roberts, Policy Analyst, Institute for Defense Analyses, Alexandria, United States

Jacqueline Simon, Norman Paterson School of International Affairs (NPSIA), Carleton University, Ottawa, Canada

Peter Stibrany, Iconitec Multimedia Inc., Vaudreuil-Dorion, Canada

René Unger, Department of Political Science, Queen's University, Kingston, Canada

Gordon K. Vachon, Head of Verification Research, Non-Proliferation, Arms Control and Disarmament Division, Department of Foreign Affairs and International Trade, Ottawa, Canada

Gustavo Zlauvinen, Senior Political Advisor, United Nations Special Commission (UNSCOM), United Nations, New York, United States

Editorial Foreword

The chapters included in this volume are based on papers commissioned for and presented at the Fourteenth Annual Ottawa NACD Verification Symposium, sponsored by the Non-Proliferation, Arms Control and Disarmament Division of the Department of Foreign Affairs and International Trade (Canada) and organised by the York Centre for International and Security Studies. The Symposium was held in Montebello, Québec, 12-15 March 1997. A group of 40 experts from government, the military, industry, and academe attended the conference. The volume generally reflects the sequence of presentations made during the Symposium.

Using the unprecedented success achieved in multilateral non-proliferation, arms control, and disarmament (NACD) negotiations since 1990 (CFE, CWC, CTBT, extended NPT) as a background, participants at the Symposium explored practical problems inherent in the implementation of agreements. Two issue areas, cyberspace and outer space, were examined in an effort to determine what implications each may have for the NACD process.

In the keynote address, Gustavo Zlauvinen outlines, with respect to NACD processes, the dichotomy of challenges and opportunities arising from the propagation of new technologies. While noting that technological advances often have direct military applications, Zlauvinen draws attention to some ways in which they may be harnessed for NACD purposes. However, he emphasises that new capabilities in the realms of cyberspace and outer space are not, in and of themselves, a panacea for the problems associated with NACD verification efforts. Rather, Zlauvinen argues for a more balanced approach to NACD verification wherein new technologies — perhaps managed and employed under UN auspices as a supplement to national efforts — are used as enhancements to more traditional means, measures, and strategies undertaken by the parties to NACD agreements.

In Chapter One, Ambassador Jonathan Dean reflects upon the myriad NACD successes in the decade from 1986 to 1996 and poses the question of whether the spirit and momentum of these achievements is sustainable. Given the prevailing political conditions in Russia and the United States, coupled with the increasingly staunch demands by countries such as India for commitment to an unambiguous timetable for the elimination of nuclear weapons, his rather bleak assessment is that it seems unlikely to be. Nevertheless, Dean suggests a range of opportunities which may yet exist to breathe new life into faltering NACD processes.

Ambassador Thomas Graham, Jr. then outlines what he argues ought to be the priorities on the arms control agenda in the immediate term. Here, he emphasises a strengthened NPT review process, enhancement of the BWC, and the expedited ratification and implementation of START II, the CTBT, and the CWC. Graham also draws attention to the possibility that the independence of the United States Arms Control and Disarmament Agency (ACDA) may be abolished — a possibility which he argues may have dire consequences for the future of NACD efforts.

In Chapter Three, Patricia Bliss McFate considers some new approaches to the evaluation, verification, and implementation of an ever-increasing number of NACD agreements, measures, and regimes. She argues that an opportune, and yet essentially unexploited, means by which to enhance existing NACD processes is via the harmonisation of the various regimes. In this connection, periodic consultations, shared intelligence, a concerted effort to marshal support from the international community for enforcement of global norms of non-proliferation, and, where appropriate, more inclusive consultation with potential recipients are suggested as modest ways to implement harmonisation.

In Chapter Four, Richard Guthrie presents us with an overview of what is happening in the field of non-proliferation, arms control, and disarmament, with a particular focus on what may be the significant issues of the next few years. Guthrie discusses regional treaty regimes, as well as global treaty regimes, with particular emphasis on the benefits of universality.

Gordon K. Vachon explores the relationship between multilateral export control regimes and multilateral arms control regimes in Chapter Five. He considers the tensions between national responsibility and international organisational responsibility as bearers of challenges for multilateralism in the coming years.

In Chapter Six, F.R. (Ron) Cleminson provides a brief overview of the outer space issue as the 21st century approaches and identifies certain changes and activities relevant to the principles of the non-weaponisation of space. Cleminson notes that most states, Canada among them, support the notion that outer space should be maintained as a weapons-free environment. However, according to Cleminson, space-based remote sensing remains an important part of a monitoring and verification package for present and future NACD agreements. Like Zlauvinen, Cleminson suggests a possible role for the United Nations in this regard.

The next three chapters inquire into cyberspace and the Information Revolution and their relationship to non-proliferation, arms control, and verification issues. In Chapter Seven, David Mussington provides us with an analysis of the emergence and structure of cyberspace. He sees cyberspace as variously a channel for the flow of proscribed information relating to military technology, itself a target of hostile action (information warfare), and as a potential new verification tool. Jeffrey R. Cooper and Christopher Burton follow this with a discussion of some of the implications of the Information Revolution, both in the broader realm of national security and in the area of arms control efforts in particular. They conclude that, since information technologies can be neither effectively controlled nor restrained, we must understand and accept the new vulnerabilities which they may engender, even as we exploit their potential to help overcome others. Finally, James Der Derian takes us on a cleverly conceived 'tour' ('cut and paste', 'buzzcut forward', 'flash sidewise') through the very sources and implications of the new technologies which he seeks to highlight.

In Chapter Ten, Brad Roberts offers us some speculative thinking on what role arms control can play in meeting the security challenges of the emerging strategic era. He draws particular attention to the problem of uncertainty in the United States as to the future utility of arms control *vis à vis* national and international security. With this in mind, Roberts argues for the necessity of rethinking old assumptions about arms control and for evaluation of the new dynamics of international security. To this end, he undertakes to evaluate the utility of arms control with respect to several specific world order tasks.

The next four chapters all examine various aspects of verification, in particular, aerial surveillance and overhead imagery. Alan Crawford, in Chapter Eleven, argues that there is a legitimate and practical role for the United Nations in certain specific and carefully delimited areas of verification, with special emphasis on what Canada can and should do in support of this role. René Unger follows this with a comparison of the aerial surveillance methods and technologies used in the Sinai Field Mission, the Multinational Force and Observers (MFO) which followed it, and the United Nations Special Commission on Iraq (UNSCOM). Unger suggests that there are commonalities between such issues as cost, the degree of interaction between aerial surveillance activities and other verification efforts, the complexity of sensors required, the relative merits of high- versus low-altitude surveillance, and the procedures needed to assess information gleaned from aerial surveillance. In Chapter Thirteen, Phillip J. Baines focuses on the opportunities to be had from commercially available high-resolution, spaceborne imagery in an era when national governments increasingly seek the means for effective verification at lower levels of expenditure. Finally, Peter Stibrany provides us with a summary of a workshop on the use of satellite imagery in the verification of arms control agreements, concentrating specifically on the Comprehensive Test Ban Treaty.

In Chapter Fifteen, Jason Cameron examines efforts by the International Atomic Energy Agency (IAEA) to develop a strengthened comprehensive safeguards system. Cameron looks at the "93 + 2" programme to determine if significant steps have been taken toward strengthening the IAEA's ability to detect diversions of declared nuclear material as well as to detect undeclared activities.

In Chapter Sixteen, David DeClerq surveys the issue of light weapons as a new focus for arms control and disarmament. Herein he addresses the question of whether it is appropriate to address misuse from a supply-side perspective, a demand-side perspective, or both. While he concludes that rapid or even moderate progress is unlikely, he does identify ways in which at least incremental steps can be taken to mitigate against the destabilising effects of light weapons surpluses.

The next three chapters look at specific national case studies. Sergei Plekhanov begins with a survey of the implications of crisis in Russia for arms control and disarmament. In Chapter Eighteen, George Lindsey and Jim Bayer outline a pair of related studies which inquire into possible future challenges for multilateral arms control on the Korean Peninsula. This is followed by a second offering from George Lindsey which summarises the perspectives on Korean arms control through the eyes of extra-Peninsula players.

Finally, in Chapter Twenty, Jacqueline Simon furnishes an integrated summary of some key themes which recurred throughout the symposium.

We are indebted to all conference participants for their commitment and involvement, and in particular, to Gustavo Zlauvinen, Ambassador Thomas Graham, Jr., and Sergei Plekhanov, each of whom gave a dinner address. Special appreciation goes to those authors who have contributed papers for this volume, and to the Non-Proliferation, Arms Control and Disarmament Division, particularly F.R. (Ron) Cleminson, Alan Crawford, Jill Sinclair, and Gordon Vachon.

Finally, we would like to thank our colleagues at the Centre for International and Security Studies for their assistance and support, including Heather Chestnutt, David Dewitt, Wendy Kubasik, and Karen Warner. It is hoped that these papers will stimulate further interest, debate, and analysis of the issues examined herein. Of course, the editors are responsible for any errors of fact and interpretation.

J. Marshall Beier and Steven Mataija
York University, Toronto
May 1997

Keynote Address

Meeting the Multilateral Proliferation Challenge Through United Nations Actions

Gustavo Zlauvinen*

Non-proliferation, arms control, and disarmament processes are facing a growing number of challenges. Many of them are related to new technologies in the development, production, and deployment of weapon systems. Computer simulation for testing nuclear devices, computer modelling for missile design, or the development of space-based anti-missile devices are just a few examples of such a challenge.

On the other hand, there is a growing use — or a better use — of new technologies for arms control and disarmament verification activities, such as aerial imagery, sample analysis, seismic detection, real-time data, or intelligence gathering. However, one must recognise the limitations of such technologies in the field of disarmament. For example, the intelligence gathering can help in assessing the compliance of a treaty's party, but it needs to be double-checked with other means, for in many cases the information can be incorrect or biased. But even with this limitation, the more information flowing among members of an arms control treaty, the more transparent the system. Thus, it stimulates confidence-building among parties and increases the "deterrence" factor for possible violators of the agreement, as it would be more difficult for a party to violate the treaty without being detected.

There are also limitations to space-based technologies with regard to their use in disarmament verification. If the state under satellite surveillance is aware of being "observed," it can take countermeasures to cover its proscribed activities. Additionally, these types of technologies are expensive, sensitive, and available to only a handful of countries. Remote-sensing can help in detecting new buildings, troops movements, or other military activities, but the information is not always available in real-time. Thus, by the time the analysis of the information is ready, the situation may have changed.

---

* This chapter is an edited version of a presentation made to the Fourteenth Annual Ottawa NACD Verification Symposium, "Cyberspace and Outer Space: Transitional Challenges for Multilateral Verification in the 21st Century," 12–15 March 1997, Montebello, Quebec, Canada.

During the early phase of nuclear test ban negotiations during the late 1950s and the early 1960s, the state of ignorance or lack of information resting with any third party was probably similar to the one that exists with regard to space disarmament today. The science and technology of nuclear fission were very closely guarded military secrets. For example, countries other than those in possession of such weapons did not know about the practical engineering of atmospheric testing and its relationship to the advancement of weapon science and design.

However, during the last ten to twenty years, the gradual increase in the level of general knowledge and experience gained from the use of nuclear energy for peaceful purposes has formed an important basis for discussions among the nations of the world in the disarmament and non-proliferation fields. Something comparable is taking place today in the case of outer space. And the growing commercial use of space technologies is contributing to the dissemination of a better understanding of the possibilities of a destabilising militarisation of outer space. Thus, the basis for technical negotiations in multilateral fora is now available to fill the gaps in previous treaties on the prohibition of the militarisation of outer space.

The relationship between the level of knowledge and the degree of participation in the arms control and disarmament regimes is illustrated in the exchange that can occur over treaty violations. In the absence of detailed knowledge of the scope and technical definitions of any agreement, it is not possible for a third party to pass independent judgement about a possible treaty violation. Thus, it is important to establish objective, credible, and reliable verification mechanisms for arms control and disarmament treaties, with equal participation from all the parties, regardless of their different knowledge levels.

Due to the new monopoly on accurate information, in many cases the discussions of disarmament are based on data provided by a reduced number of countries. This covers almost the entirety of modern high technology, ranging from high-response, high-resolution electro-optical sensors, high-accuracy positioning, pointing and tracking devices, reliable and accurate high-density communication, and very complicated management software. But there is a recurrent unwillingness of governments to make these new technologies and the data derived from them available to other parties.

Furthermore, no arms control or disarmament measure can be fruitfully implemented unless there is a necessary minimum amount of mutual confidence established between the parties. However, in many situations this is not the case. When countries are negotiating disarmament measures with the full expectation that the other side will, immediately after signature,violate those measures, there can be no way to assure proper implementation unless there is a reliable verification mechanism.

Since its creation, the United Nations has played an important role in the negotiation of disarmament and non-proliferation agreements. However, with few exceptions, it has not been used to all its potential in the verification of such agreements. Most of these treaties have established review conferences or peer review meetings, where at the same parties of the treaties are supposed to assess the implementation of the agreements by themselves. But again, for most of the parties, with limited knowledge and expertise in such complex and highly sophisticated issues, this limitation reduces their participation in the assessment of the compliance by other parties with the treaty's obligations.

A solution to this unequal situation could be provided by tooling the United Nations with highly qualified international experts, to serve as an 'impartial' and 'separated' body of verification for such agreements. The UN could then provide technical and impartial assessments to all parties of an arms control, non-proliferation, or disarmament agreement, in order to help them in dealing with discrepancies with regard to compliance with the agreement by others. It could also help in imposing multilateral sanctions on those parties which violate the agreement.

Few, but very important examples of UN participation in similar processes can be identified in the works of the International Atomic Energy Agency (IAEA) and the Special Commission (UNSCOM).

When the Government of Iraq was defeated by Coalition forces in 1991, it accepted a set of obligations as conditions for the ceasefire. But, as the international community distrusted the Iraqi government, the Coalition forces left it to the United Nations to conduct the task of verifying the destruction of all Iraqi weapons of mass destruction and to monitor against re-acquisition of such capabilities. Consequently, the Security Council established UNSCOM as its subsidiary body for the implementation of these two tasks, in cooperation with the International Atomic Energy Agency which has the primary responsibility for nuclear weapons issues.

During its almost six years of existence, the Special Commission has acquired a unique expertise in the detection, elucidation, and elimination of major biological, chemical, and long-range missile programmes. It has also developed ongoing measures to ensure that these programmes and capabilities are not restarted by Iraq. In this process, UNSCOM has used and perfected technologies already available and, in several cases, it has developed and introduced innovative approaches and techniques. Some of these may have applicability to other non-proliferation and disarmament processes.

The Special Commission is using an array of interrelated activities such as: no-notice inspections, remote ground-based sensors, tags and seals, sample taking and analysis, and a variety of detection technologies, as well as notifications from governments of trade with Iraq in items of a dual-use nature. A high-altitude U-2 reconnaissance aircraft scans for telltale signals of suspicious activities that may necessitate on-site investigation.

As an example of the use of new technologies for disarmament verification, UNSCOM has collected samples from equipment at several facilities in Iraq which had been declared by Baghdad or suspected by the Special Commission of having been related to a biological weapons programme. The results of the sampling analysis have elicited evidence of undisclosed production of biological warfare agents by Iraq. This sampling analysis has proven to be effective even where an agent has most likely not been present for some years and where equipment had been subject to rather vigorous cleansing by Iraq to remove any trace of such activity.

However, we have to point out the limitations for such technologies. In the case of the sampling analysis, identifying and narrowing the domain to sample is perhaps the single most difficult problem. Technology has brought UNSCOM to the stage that if a micro-organism or its product is present, it can be found and identified. If, however, the sampling domain is not narrowed, the cost of sampling, both monetary and political, is prohibitive. For example, sampling the soil where biological weapons field trials were conducted in Iraq is likely to provide essential information as to what agents were subjected to trials, either confirming what Iraq has declared or identifying additional agents that may not have been declared. However, if the areas of such trials cannot be sufficiently identified and limited, then such a technique is impractical. Negative answers may only mean that the wrong area was sampled.

Another example of the use of new technologies for verification can be provided by the current 'forensic analysis' which the Special Commission is conducting on destroyed Iraqi missile engines and components to determine the circumstances of Iraq's unilateral destruction programme of 1991. This process will include:

- the material analysis of remnants to establish engine firing, identification of traces of missile fuel or combustion products on remnants;
- computer reconstitution of destroyed missile components to determine whether engine remnants without identifiable serial numbers represent additional engines or constitute only parts of engines already counted;
- material analysis of engine areas affected by explosion, to establish the methods of the unilateral destruction;
- the placement of explosives on engines or airframes, to identify traces of materials from component-to-component contact during explosion and to conduct modelling of explosive damage that could have been produced by explosive methods declared by Iraq, in order to determine whether major critical engine components have been removed from the engines prior to their destruction;
- comparative analysis of soil samples on missile remnants, to establish specific location of the destruction and burial sites; and,

- comparative study and material analysis of remnants whose origin has not yet been established, to determine whether components of Iraqi manufactured engines had been included or replaced the destroyed missiles of Soviet origin.

The forensic analysis approach is also used by UNSCOM in the determination of authenticity of documents provided by Iraq as means of verification, as well as to analyse traces of prohibited items, such as chemical and biological warfare agents or precursors.

Over the past five years, UNSCOM has made considerable progress in the identification and the elimination of Iraq's proscribed weapons and capabilities. The most notable success in the chemical weapons area has been the destruction of Iraq's war stocks of nerve gas and mustard gas, using destruction facilities constructed by the Special Commission. However, Iraq has not yet provided enough evidence of the alleged 1991 unilateral destruction of chemical agents, precursors, and chemical warheads, nor on the real production capabilities of VX, the most lethal and advanced nerve agent.

In the biological weapons area, the detection of an extraordinarily large amount of growth medium and of a large fermenter capability, as well as other signals, led UNSCOM's experts to the conclusion that Iraq had had a viable biological weapons programme. After Iraq admitted the existence of this programme in 1995, the Special Commission demolished the Al-Hakam, the main biological weapons production facility, as well as other installations and equipment related to this secret weapons programme. This was the major event in one of the most effective investigations in arms control history.

With respect to missiles, UNSCOM has supervised the destruction of 49 long-range missiles and related equipment. Its inspection activities have furthermore disclosed vigorous efforts by Iraq to develop long-range missile capabilities up to and above 1,500 miles. In 1995, UNSCOM detected a purchase of highly sophisticated guidance systems for long-range missiles. The technical structure of the longer-range missile programme indicates that delivery of biological or possible chemical weapons was the purpose of the efforts in this respect.

In spite of many setbacks, the international inspectors, organized and lead by the IAEA Action Team, have gradually reached a good understanding of the extent and scope of Iraq's nuclear programme. The most important facilities and equipment have been destroyed or demolished.

In order to monitor Iraq's dual-use capabilities, the Special Commission has developed an array of interrelated activities, such as: regular no-notice, on-site inspections; control of dual-use facilities and equipment; aerial surveillance with a variety of sensors based on five helicopters and on the U-2 aircraft; tags and seals on dual-use equipment and short-range missiles; remote ground-based sensors; sample taking and analysis; information obtained from other sources; chemical and biological laboratories; hundreds of cameras sending real-time photos of Iraq's production installations; and, a special export/import team working on data provided by governments on their exports to Iraq in order to monitor against dual-use items. No one of these elements on its own would suffice to provide confidence in the system, but together they constitute the most comprehensive international monitoring system ever established in the sphere of arms control.

But despite all these efforts, UNSCOM has still not reached the stage where it can state with confidence that everything that is proscribed to Iraq has been identified and eliminated. It continues to believe that limited, but highly significant quantities may remain, as Iraq has not been able to account for a number of proscribed missiles and certain high quality chemical and biological warfare agents and related capabilities which it had acquired. But probably the most important difficulty experienced by UNSCOM is the fact that its activities are conducted in a 'hostile' environment. The Special Commission's mission is one of enforcement and its inspectors are working in an environment of opposition, counter-intelligence, denials, deceit, and psychological warfare. And unlike other disarmament mechanisms, wherein the parties voluntarily accept international inspections and can even negotiate them, UNSCOM's rights to inspect and monitor are imposed on Iraq against its will. As a punishment for Iraq's aggression, the Security Council decided that the oil embargo on Iraq would remain in force until that body could agree that Iraq's obligations with regard to the elimination of all its weapons of mass destruction had been implemented. Thus, a powerful incentive for Iraq to accept the dismantling and disposal of its mass destruction capabilities was to offer it the prospect of quickly getting access to the revenues from its oil exports.

However, the Iraqi leadership has chosen to try to retain the prohibited weapons. A safe assumption is that Iraq's policy is both to retain the prohibited weapons and to achieve, at the same time, an end to the oil embargo through the dispensation of promises of generous compensation to those who do not line up against it. A breakup of Iraq's political isolation is no doubt part of the calculations behind the contracts for oil exploration in Iraq which have been reported in recent months.

However, following disclosures by Iraq and findings by UNSCOM after the defection of Hussein Kamel, Saddam Hussein's son-in-law, in 1995, the Special Commission has obtained a fairly good picture of how the security forces, not the army, are hiding and moving prohibited weapons and equipment. Through a series of inspections, supported by our U-2 aircraft, we have put the concealment system under considerable stress. As the institutions hiding the prohibited items also are responsible for the security of the leadership, considerable friction and tension has emerged between the inspectors and the Iraqi security apparatus. The Special Commission will continue to use all its capabilities to uncover and destroy any hidden prohibited weapon, item, or component in Iraq.

For the Special Commission, the Iraqi policy has radically changed the expectations which were in existence at the outset of its work. Instead of being an agent for technical verification of disarmament activities under essentially cooperative circumstances, it has been obliged to transform itself into an investigative and enforcement agency, confronting blockage, cheating, denials, and threats with a complex political and security agenda to attend. Only if UNSCOM and the IAEA are allowed to carry out their work, with the full support of the Security Council, will the UN be able to contain threats from Iraq.

A notable part of the decision to end the Gulf War was the handing over to the United Nations of the responsibility for the execution of the ceasefire. The cornerstone of these efforts has been the right of UNSCOM and the IAEA to conduct rigorous, no-notice, on-site inspections anywhere in Iraq — the only such inspections ever by a United Nations body and the most intrusive ever devised in the field of international arms control.

The success of the UN activities in Iraq have encouraged disarmament negotiators to press for more intrusive inspections in several areas, from strengthening the IAEA safeguards system — the so-called "93+2" Programme — to new verification inspections for the 1972 Biological and Toxin Weapons Convention, to regional agreements such as the African Nuclear Weapons Free Zone. Therefore, we can almost predict that any further participation of the United Nations as a tool for future disarmament and non-proliferation processes will depend, to an important extent, on how effectively the Special Commission and the IAEA have shouldered the responsibilities entrusted upon them thus far.

Chapter One

Where Are We Now; Where Are We Going in Arms Control?

Jonathan Dean*

Over the past year, the question that worried many of us at previous symposia has become more insistent: whether a decade of unparalleled progress in arms control and disarmament, between 1986 and 1996, is coming to a close and, if so, what we can do about it. It seems that we no longer have the impetus and the leadership that brought forth the Stockholm Document on Confidence-Building, the Intermediate-Range Nuclear Forces (INF) Treaty, the Two-Plus-Four Treaty, the Treaty on Conventional Armed Forces in Europe (CFE), START I, START II, the brilliant diplomacy of the Lisbon Protocols, the Open Skies Treaty — to which Canada made an especially important contribution — the Chemical Weapons Convention (CWC), the indefinite extension of the Nuclear Non-Proliferation Treaty (NPT) — where Canada also played a very active role — and the signature of the Comprehensive Test Ban Treaty (CTBT) last year.

In the United States Senate, ratification of the Chemical Weapons Convention may or may not slip through. The Senate is also unhappy with last year's modifications of the CFE Treaty — which it considers treaty amendments that must be ratified — and with the Comprehensive Test Ban Treaty, which will be up for ratification soon. The Senate Republican leadership is preoccupied with domestic policy, the Chairman of the Senate Foreign Relations Committee is a tough-bargaining isolationist, and the Clinton Administration remains unwilling to abandon bipartisanship to fight the Congress, even on issues which the public supports and where its leadership would count — like paying United Nations dues. Moreover, the Administration is in serious trouble on election financing and the mid-term elections next year may increase Republican majorities.

For its part, Russia has weak leadership at a critical juncture. The March 1997 Helsinki summit may bring the beginnings of partial Russian acquies-

---

* This chapter is an edited version of a presentation made to the Fourteenth Annual Ottawa NACD Verification Symposium, "Cyberspace and Outer Space: Transitional Challenges for Multilateral Verification in the 21st Century," 12–15 March 1997, Montebello, Quebec, Canada.

cence to NATO enlargement. But it is very doubtful that there will be full agreement on the whole of the agenda of the summit. No matter what happens at Helsinki, NATO's open-ended invitation last December to other members of the Partnership for Peace to apply for full NATO membership guarantees that eight or more European countries will energetically push for full membership over the next decade, reinforcing resentful Russian nationalism to a dangerous extent.

I do not foresee rapid START ratification in these circumstances, even though the administration is showing considerable flexibility with regard to START III and further modifications of the CFE Treaty. In the latter part of the year, the unratified state of START II could become a very neuralgic issue in U.S.-Russian relations and East-West relations generally.

If START II is ultimately ratified, START III would, of course, provide an agenda for further action. However, even if Yeltsin recovers some of the vigour and energy that has been absent the past eight months, the Russian national leadership is preoccupied with national survival in the fundamental sense. We may be pleasantly surprised on this point, but there will probably not be a Russian leader in the next decade with strong positive interest in disarmament and the authority to actively push through a START III programme, and to go beyond it — or a Russian public that will actively indicate its own interest in this subject. This development is especially unfortunate because the end of the Cold War is at last having a predictable impact on Western expert opinion with regard to nuclear weapons and nuclear disarmament.

In recent months, an increasing number of respected individuals and organisations have joined in advocating the complete elimination of nuclear weapons within the foreseeable future, in most cases without specifying a specific time limit. These include the Atlantic Council; the Stimson Center; the Canberra Commission; as well as Nobel Prize co-recipients Pugwash Conferences and Dr. Joseph Rotblat; General Lee Butler, former commander of the U.S. Strategic Air Command; former NATO commander General Andrew Goodpaster; and, a group of 61 retired generals and admirals from seventeen countries.

In Washington, a new coalition has been founded — the Committee on Nuclear Policy — that pulls together ten or more projects which advocate radical cuts in nuclear forces. One of them, the Deep Cuts Project, advocates that all five nuclear weapons states (NWS) reduce force levels to largely immobilized, internationally monitored arsenals of 200 warheads each. The project aims to use the leverage from NWS agreement to make deep cuts to bring the threshold countries to accept a fissile material cut-off and to place their nuclear materials and delivery systems in monitored storage like the nuclear weapons states. The advantage of these programmes is that increasingly they are tackling the practical problems of deep cuts, such as verification and the survivability of small arsenals. Their disadvantage is that they lack a sponsor. This enthusiasm for far-reaching action is not reciprocated by Russia or by the Clinton Administration. Public interest in this subject is low as long as there is no immediate threat.

On the multilateral front, the negotiation for a fissile material cut-off treaty is blocked in the Conference on Disarmament (CD) and India has become a more aggressive blocking player with its demands for time-bound commitments on elimination of nuclear weapons.

What can we do to counteract these negative trends? First, and most immediate, we need to devise an acceptable overall concept for a possible second stage of NATO enlargement; as the second stage is currently conceived, it could have extremely dire consequences.

Second, we appear to be entering a period when the nuclear weapons states will be asked more insistently than ever before to specify their preconditions for eliminating nuclear weapons. Since debate on this subject flickered out at the outset of the Cold War nuclear confrontation, there has been relatively little sustained effort either by NWS officials or those urging abolition of nuclear weapons to identify the conditions for total elimination of nuclear weapons. Much more public discussion of this subject is needed. Officials of NWS governments should be reminded of their NPT obligation to eliminate their nuclear arsenals and asked to specify the conditions under which they will be prepared for decisive action. Whatever preconditions they present, these should be analysed to determine whether they have some substance or are merely excuses for maintaining the status quo. Where they are shown to be excuses, they can be refuted and the debate over abolition will be simplified. If some preconditions — like bolstering regional security organisations and improving worldwide conflict prevention and peacekeeping — are shown to have objective justification, they could become common goals of NWS governments as well as those pressing for elimination of all nuclear weapons.

A third area of potential progress, even in these difficult conditions, is 'de-alerting'. At the urging of the Canberra Commission and many non-governmental organisations (NGOs), governments are looking more closely at agreed de-alerting. However, up to now, the Clinton Administration has viewed de-alerting as tied to agreed force reductions. There may also be difficulty in persuading Russia to de-alert on a large scale. What is needed is a standardized, mutually verifiable system which is rapidly reversible. The two most plausible candidates — removal of missile nosecones/shrouds or dismantling warheads from missiles — have the disadvantage of requiring missile submarines or mobile missiles to cluster vulnerably for dismounting or remounting nosecones or warheads. Both actions can be made verifiable, but take a long time to make retroactive — especially dismantling warheads. We have to identify a practical form of de-alerting.

A benefit here is that this kind of action could be carried out by the Russians as well as the U.S. government by executive order and would not, therefore, require a treaty. There is a challenge here for those interested in verification: for the governments of non-nuclear weapons states (NNWS) and for NGOs, the problem has been that information on subjects like warhead mounting has been highly classified because it is believed to reveal vulnerability.

Fourth, the NPT review process will begin in 1997. Governments of non-nuclear weapons states can pick up these themes and advance them there. The NWS will be pressed by the non-aligned countries to accept a time-bound commitment to eliminate their weapons. The NWS governments will resist these pressures, but they may lead to incremental moves. Some of these incremental possibilities are: a U.S.-Russian commitment to dismantle all or a fixed high percentage of the warheads withdrawn under the START treaties from operational deployment and to transfer their fissile content to bilateral storage, in order to begin a process of genuine disarmament; agreement among all NWS to research workable methods of de-alerting; agreement not to increase the number of their operationally deployed weapons; agreement among the NWS to exchange a wide range of information on their holdings of warheads and fissile materials, and themselves to declare a moratorium on production of fissile materials for weapons and to begin to negotiate among themselves a fissile material cut-off treaty for submission to the Conference on Disarmament.

It is my hope that a group of non-nuclear weapon states — among them Canada, Germany, and Japan — will create an informal coalition, that they will insistently ask the NWS in this and succeeding Prepcoms to clarify their preconditions for eliminating their weapons, and that they will promote the same programme of incremental steps in successive Prepcoms, helping the NPT review process to realize its real potential as an engine for nuclear disarmament.

Finally, there is the area of conventional arms. The progress toward a ban on anti-personnel landmines, sparked by Canada, is a bright spot. But, with the exception of the CFE and landmines, the capacity of the international community to cope with the problem of conventional arms and conflicts fought with conventional arms throughout the world has been inadequate. The grim statistics illustrate this — 45 million dead since World War II and 30 major wars now in progress. One useful short-term measure would be to make a review of the worldwide situation of conventional disarmament a standard part of the NPT review process — perhaps one or more regions could be analysed in each Prepcom, with attention to the global situation in the review conference itself. I believe this idea may be taken up at the Prepcom. Hopefully it will not become viewed by non-nuclear weapons states as a 'smokescreen' by the NWS for inaction on nuclear cuts.

One major difficulty in the fields of conventional disarmament and of peacekeeping is that they lack an integrating general concept, a vehicle for focusing the attention of a larger number of governments over a long period. A working group from the Washington area United Nations Association has decided to tackle this problem by combining conventional disarmament, peacekeeping, and conflict prevention in a single global programme aimed at lowering the frequency of armed conflict.[1] This group hopes to raise some interest in this project in the Disarmament Commission and the Conference on Disarmament.

1. See the Report by the Task Force on Peace and Security, United Nations Association of the USA, National Capital Area, *A Global Treaty for Reducing Conventional Arms and Armed Conflict*, 1997.

Chapter Two

The 1997 Multilateral Arms Control Agenda and ACDA Priorities

Thomas Graham, Jr.*

The last few years have been a time of unprecedented success in arms control. We have seen the cornerstone of United States and Canadian efforts to prevent the spread of nuclear weapons — the Nuclear Non-Proliferation Treaty (NPT) — indefinitely extended. A significant step toward reducing the over-armament of the Cold War was taken when START II was ratified by the U.S. Senate. And the signing of the Comprehensive Test Ban Treaty (CTBT) ended the era of nuclear testing, a goal that had been sought for more than forty years.

Despite this string of successes, however, the world remains a dangerous place. The spread of technology has made it possible for dozens of states — many of whom view the United States as an enemy, or whose interests are inimical to those of the United States — to possess or seek nuclear, biological, or chemical weapons capability.

The collapse of the Soviet Union has created new fears about nuclear materials leaking out to terrorists or criminals, and the efforts of 'rogue' states such as Iran and Iraq to acquire weapons of mass destruction has become more threatening in recent years. We know now how close Iraq was to building a nuclear weapon just prior to the Gulf War, and Saddam Hussein continues to hide information concerning the Iraqi chemical, biological, and nuclear weapons programmes from UN inspectors. In Libya, construction has ceased for the time being on what is thought to be a massive chemical weapons plant at Tarhunah. These states, and several others, who have time and again shown that they operate outside the circle of 'civilised' nations by supporting terrorism, must be prevented from acquiring weapons of mass destruction.

Clearly, much more remains to be done to reduce the threat posed to peace and security by weapons of mass destruction. As U.S. President Clinton has noted, we are engaged in "the most ambitious agenda to dismantle and fight

---

* This chapter is an edited version of a presentation made to the Fourteenth Annual Ottawa NACD Verification Symposium, "Cyberspace and Outer Space: Transitional Challenges for Multilateral Verification in the 21st Century," 12–15 March 1997, Montebello, Quebec, Canada.

the spread of weapons of mass destruction since the dawn of the nuclear age." On this agenda are several important items that will reduce this threat and, hopefully, the next few years will bring as much success and security as have the last few.

The primary areas of concern in the next year will be the ratification and strengthening of existing treaties. We expect and hope that the Russian Duma will soon ratify START II, thus bringing the strategic nuclear arsenals of the U.S. and Russia down two-thirds from their Cold War high. Decisions must be made on how to effectively implement the strengthened NPT review process and negotiations must also continue on how to enhance the Biological Weapons Convention (BWC). It is also vital that the ratification process of the CTBT move forward so as to bring this landmark treaty into force as soon as possible. Perhaps most important, the Chemical Weapons Convention (CWC) must be ratified by the United States soon to ensure continued U.S. leadership in the fight against the spread of chemical weapons.

START II

Although the Soviet Union has collapsed and the United States no longer views Russia as an enemy, the huge nuclear arsenals left over from the Cold War pose a potentially serious threat. The United States and Russia have attempted to lessen this threat through the negotiation of bilateral disarmament treaties such as START I and START II. The START I Treaty, which mandated reductions in the total number of deployed strategic warheads to 6,000 on each side — roughly a one-third cut — entered into force on 5 December 1994. The United States Senate gave its advice and consent to the START II Treaty in January 1996. Unfortunately, over a year later, we are still waiting for the Russian Duma to follow suit. Building on the disarmament progress made under START I, START II will leave each side with 3,500 deployed strategic warheads. Entry-into-force of START II will also facilitate negotiation of further reductions — leading perhaps to a START III. President Clinton reaffirmed this commitment last year in his speech before the United Nations General Assembly, when he stated that "When Russia ratifies START II, President Yeltsin and I are ready to discuss the possibilities of further cuts." The President also called for the limiting and monitoring of nuclear warheads and materials to help make deep reductions irreversible. Although the Russians have concerns about NATO expansion and U.S. commitment to the ABM Treaty, I am optimistic that they will recognise their security interests are best served by ratifying START II.

Comprehensive Test BanTreaty

A Comprehensive Test Ban Treaty is a bulwark against the spread and further development of nuclear weapons capabilities. The CTBT places a profound and permanent new constraint upon nuclear weapons capabilities, particularly on the nuclear weapons states. It will constrain any nation from improving its existing arsenal and prevents the development of a new generation of nuclear weapons. It also keeps new states from becoming nuclear powers by preventing them from testing in order to learn how to build nuclear weapons more efficiently, or to make more advanced weapons. Both of these results ensure that the arms race is over once and for all and are essential prerequisites to further progress toward nuclear disarmament. As President Clinton recently remarked, the CTBT points us "toward a century in which the roles and risks of nuclear weapons can be further reduced, and ultimately eliminated."

Now that the CTBT has been opened for signature and 142 countries have signed the treaty, we must begin work to secure ratification by the required parties to bring the treaty into force. The goal of the Clinton Administration is to work toward achieving entry-into-force of the CTBT at the earliest possible date: September 1998. Of the 44 countries whose ratification is necessary for entry-into-force, 41 have already signed the treaty, including all five of the declared nuclear weapons states, as well as Israel. While it is obviously of extreme importance that the remaining three countries — India, Pakistan, and North Korea — sign and ratify the CTBT, it is crucial that the other signatories begin the ratification process as well. A strong international consensus against nuclear explosive testing already exists, but each signature and ratification serves to further codify this international norm and make it stronger. In the United States, the CTBT likely will be submitted to the Senate in a few months, and although the treaty will be subject to serious debate, I am confident that the Senate will give the CTBT its advice and consent to ratification.

Nuclear Non-Proliferation Treaty

The indefinite extension of the Nuclear Non-Proliferation Treaty (NPT) in May 1995 was a significant step forward in fighting the proliferation of nuclear weapons. With 185 countries Party to the now-permanent NPT, it is clear that the international community has taken a stand against the further spread of nuclear weapons. Before the NPT entered into force in 1970, the acquisition of nuclear weapons capability had often been a point of national pride. The NPT has made it tantamount to a violation of international law and has added immeasurably to the security of the United States, Canada, and of the entire world.

Since the NPT entered into force in 1970, periodic Review Conferences (RevCon) have been held, which provide Parties with the opportunity to review the operation of the Treaty to ensure that its preamble and provisions are being realised and observed. The next RevCon will take place in the year 2000. The first Preparatory Committee (PrepCom) meeting leading to the 2000 NPT Review Conference will be held in April 1997. This is the first PrepCom under the post-1995 NPT Conference regime and it will be an important step toward the "strengthened treaty review process" called for at the 1995 NPT Review and Extension Conference. The meeting will address exactly how the "strengthened treaty review process" should be implemented.

The PrepCom will make both the substantive and procedural preparations for the 2000 NPT Review Conference. Over the next several weeks, attention will be increasingly turned toward this issue as States Parties begin their assessment of how to strengthen the NPT regime. The United States is looking forward to a constructive, cooperative process leading to the 2000 NPT Review Conference.

Biological and Toxin Weapons Convention

The Biological and Toxin Weapons Convention (BTWC) entered into force in 1975, and has 138 Parties who have pledged never under any circumstances to develop, produce, stockpile, or otherwise acquire or retain microbial or other biological agents, or toxins — whatever their origin or method of production — of types and in quantities that have no justification for prophylactic, protective, or other peaceful purposes. The United States has passed legislation making violation of the BTWC a criminal offense, and the U.S. Anti-terrorism Act of 1996 improves control over potential biological warfare agents. However, unlike other regimes, the BTWC contains no provisions for on-site inspection activity — a source of increasing criticism in recent years. In an effort to strengthen the BTWC, many of the Parties are currently actively participating in an effort to draft a legally binding protocol that will enhance openness and transparency and thereby, improve compliance. This instrument will hopefully set forth measures that provide for off-site and on-site inspection activities and should strengthen compliance by making certain national information declarations mandatory. By 1998, well before the Fifth Review Conference in 2001, we hope to achieve such a legally binding instrument in the form of a Protocol to the BTWC.

Chemical Weapons Convention

Last, but definitely not least, we come to the Chemical Weapons Convention (CWC) of 1993, which will ban the stockpile, transfer, and production of chemical weapons; eliminate stockpiles now in existence; and, require parties to submit to intrusive on-site inspections. Due to its comprehensive verification regime, this treaty, which was drafted in consultation with representatives from the U.S. chemical industry, is a landmark in the struggle against the proliferation of weapons of mass destruction. The Convention will make it tougher for 'rogue' states to acquire chemical weapons and will reduce the threat to American citizens at home as well as U.S. troops in the field.

With more than the 65 states necessary to trigger the 180-day countdown toward entry-into-force now having ratified the CWC, the Convention will enter into force at the end of April 1997. Certain individuals on Capitol Hill in Washington are not fond of this treaty, and, as a result, the United States has not yet ratified the CWC. Despite their opposition, President Clinton has vowed that the U.S. "will join the ranks of nations determined to prevent the spread of chemical weapons." Although the United States has unilaterally renounced the use of chemical weapons and is destroying its own chemical weapons stockpiles, failure to ratify the CWC would have disastrous consequences. The CWC was specifically designed to punish those countries which choose to flout the international norm against chemical weapons, thus providing additional incentives to join the regime. They will be subject to restrictions on their chemical trade — restrictions which will also apply to the United States chemical industry if the U.S. does not become a part of the CWC regime.

Aside from the hundreds of millions of dollars in sales and thousands of jobs the United States would lose from CWC parties being forced to apply trade restrictions to chemicals that originate in or are being shipped to the United States, American leadership in the international non-proliferation regime would be undermined. Its credibility in the fight to prevent the spread of weapons of mass destruction would be eroded were the United States to fail to join the CWC regime. In the words of retired U.S. Army General Norman Schwarzkopf, "by not ratifying the treaty, we align ourselves with nations like Libya and North Korea, and I'd just as soon not be associated with those thugs ..." If the United States does not join the CWC, it would, in effect, be legitimising the possession of chemical weapons by such states.

One of the most oft-heard criticisms of the CWC in Washington is that 'rogue' states have no intention of signing the treaty, and therefore, the United States should not either. This is a most unconvincing argument. The United States is already committed, under laws passed by Congress in 1985 and 1992, to destroying its chemical weapons. As Chairman of the U.S. Joint Chiefs of Staff, General John Shalikashvili testified, "Desert Storm proved that retaliation in kind is not required to deter the use of chemical weapons ... Our ability to deter the use of chemical weapons in a post-Cold War world will be predicated upon a robust chemical weapons-defense protection program and the ability to rapidly bring to bear superior and overwhelming force in retaliation against a chemical attack ..." Since the U.S. has already decided to destroy its own chemical weapons, the rationale that we should refrain from ratifying the CWC because a few nations may continue to pursue them is illogical. The CWC will make it harder for those countries to acquire the ingredients they need for chemical weapons.

The Convention will also require those states which become party to it to adopt domestic legislation making the manufacture or possession of chemical weapons illegal. Currently, many states lack such laws. It is worth pointing out that within weeks of the sarin gas attack on the Tokyo subway, Japan quickly ratified the CWC and approved accompanying domestic legislation. The U.S. Administration's proposed CWC implementing legislation, which must accompany ratification of the Convention, will significantly improve the United States' ability to investigate and prosecute those who try to produce chemical weapons.

The U.S. Arms Control and Disarmament Agency

Unfortunately, in order to meet these priorities, we must simultaneously pursue another, most unwanted mission: fending off attempts to abolish the independent arms control advocacy voice of the U.S. Arms Control and Disarmament Agency (ACDA), the agency created and entrusted to negotiate arms control agreements.

The original legislation of the Arms Control and Disarmament Act that created ACDA stated that "the horrors of nuclear warfare, combined with the existence of the United Nations to focus world opinion on arms control and disarmament and to facilitate negotiation, assure that the United States will be increasingly concerned with problems of the limitation, reduction, and control of armaments." Thirty-five years later, the world has changed drastically, but arms control has become an even more important means of strengthening our national security. In the words of former U.S. Secretary of Defense Perry, arms control is "defense by other means." Every weapon we take or keep out of the hands of a potential enemy through negotiation is one we do not have to defend against. Every agreement reached helps replace international security autarky with security based on international norms, rule of law, and common understanding.

United States Vice President Gore concluded two years ago that there were some areas of overlap and duplication among the U.S. Department of State, ACDA, the U.S. Agency for International Development (USAID), and the U.S. Information Agency (USIA), where efficiencies in operations and programme cost savings could be achieved. The President's Management Council also reviewed the structure of all government agencies operating overseas in order to increase effectiveness and reduce costs, and the four foreign affairs agencies were instructed to establish common administrative services. A number of actions have already been undertaken pursuant to these instructions: under the President's reform plan, the agencies have reduced staff by 4,700 positions, cut bureaucratic layers, and established common administrative services, all resulting in more than $500 million in cost savings.

Despite these actions, some on Capitol Hill remained determined to eliminate the three agencies and last year passed legislation they claimed would "revitalise" foreign affairs by abolishing ACDA, USAID, and USIA. President Clinton vetoed the Act, noting that the three agencies had already streamlined to save money, and stated that a consolidation would "damage our ability to assure the future security and prosperity of the American people."

The issue of consolidation has recently resurfaced, partly due to the controversy surrounding the CWC and partly due to U.S. Secretary of State Albright's appropriate comment during her confirmation hearing that she had an "open mind" on the issue. Some people have construed this as meaning she would not oppose efforts to eliminate ACDA, USAID, and USIA. However, Secretary Albright also stated that, "We need a structure which best serves our foreign policy interests. We also need to assure maximum value to the American taxpayer. I am confident that we can do both." In my own opinion, the optimal way to do that is to keep ACDA as a strong, vital, and independent organisation.

An independent advocate for arms control in the United States government is important. When the President must make a decision on important matters concerning national security, it is in his best interest and the best interest of the United States that he hear as many options as possible. An independent ACDA keeps arms control high and visible on the nation's agenda. Were ACDA to be completely merged with the State Department, arms control and non-proliferation would be forced to compete with many other foreign policy issues, and would necessarily get less attention at senior levels at a time when it needs more. The dominant day-to-day mission of the State Department is country relations — an entirely legitimate and necessary responsibility — of which arms control and disarmament issues are only one part. Arms control may not always win out over these broader foreign policy considerations, but since it deals with such serious issues as nuclear weapons limitation and the proliferation of weapons of mass destruction, it deserves to be heard and taken into consideration in the formulation of national security policy.

It should also be pointed out that a lean ACDA, with less than 250 employees and a budget of less than $50 million, is a national security bargain. The fact that approximately $100 billion has been saved by the implementation of arms control treaties already negotiated by ACDA makes this agency one of the most effective uses of tax dollars in the history of the United States.

Those who would eliminate the independent voice of ACDA either do not understand that arms control is a valuable tool for strengthening U.S. security or have chosen to subordinate peace and security to their own ideological goals. Claiming ACDA to be a "Cold War relic" when our agenda and our missions are, in fact, greater than they have ever been, is, at best, misleading. As Secretary Albright commented during her Senate confirmation hearing, "Arms control and non-proliferation remain a vital element in our foreign policy framework." The only people who would benefit from the elimination of the arms control advocacy of ACDA are those who believe the United States should succumb to a "bunker mentality" — that we should arm ourselves "to the teeth" and forego all communication or cooperation with other nations. Fortunately, the American people do not subscribe to these ideas.

The United States will, in the next few years, address arms control missions that will fundamentally shape its own and others' security for decades to come. Those that I have described here represent only the 'tip of the iceberg'. I believe that the optimal way to accomplish these missions is with an independent agency dedicated to their success. However, whether or not ACDA endures, arms control will continue to play an important part in U.S. defence and foreign policy. I hope that these next years are as successful as the last few have been. If so, then, not only Canada and the United States, but the rest of the world, will be a safer place.

Chapter Three

The Interface Between Treaties and Regimes: Challenges for Evaluation, Verification, and Implementation

Patricia Bliss McFate*

As the year 2000 approaches, corporations and institutions have begun to focus on how they must accommodate themselves to the advent of a new millennium. Will computers recognize dates which include the year 2000? What happens to the words "Twentieth Century" in various global enterprises, such as 20th Century Fox? While these identity crises may appear inconsequential to the general public, their resolution will have major economic costs and effects: according to the U.S. Office of Management and Budget, reprogramming of the federal government's older computers will cost $2.3 billion; the worldwide "repair bill" for finding the year 2000 problem for older computers, fixing the problem, and testing the solutions or replacing old systems with new ones has been estimated at more than $600 billion. Nomenclature and dates are only part of the changes in the market place which will be required by the advent of the 21st century: the impact of globalisation and the arrival of the 'Information Revolution' have appeared on many agendas, including the sessions at the recent World Economic Forum in Davos.

The field of non-proliferation, arms control, and disarmament (NACD) is another institution — or more properly, a set of institutions — which faces the year 2000 with a naming problem. What will we call the period beginning in 2000? Many argue that the term post-Cold War has served its purpose — to underline the fact that the East-West enmity and division has ended. Some Russians, responding to NATO expansion and U.S. proposals for robust ballistic missile defence systems, have spoken of a "Cold Peace." Thus far, the eloquent wordsmith U.S. Secretary of State Madeleine Albright has not found the right term, although she has compared the late 1990s to what Dean Acheson labeled the "new era" fifty years ago. Beyond the question of an appropriate name, the new century permits — indeed, the Information Revo-

* This chapter is an edited version of a presentation made to the Fourteenth Annual Ottawa NACD Verification Symposium, "Cyberspace and Outer Space: Transitional Challenges for Multilateral Verification in the 21st Century," 12-15 March 1997, Montebello, Quebec, Canada.

lution and the advent of globalisation call for — new thinking concerning the evaluation, verification, and implementation of an ever-increasing number of NACD agreements, measures, and regimes.[1]

The Expanding Universe of NACD Efforts

Diplomatic efforts to constrain or prevent proliferation of weapons of mass destruction (WMD), advanced delivery systems, and advanced conventional weapons have taken many shapes and formats, among them: bilateral, regional, and international arms control treaties and confidence-building measures; peace operations; national and international sanctions; supplier regimes; national and multinational export controls; security assurances; programmes to strengthen implementing bodies; transparency measures; and the development of advanced national, regional, and international technical and intelligence means for purposes of verification of obligations to control arms, reduce conflict, and prevent proliferation. Many efforts have been in response to recent historical events, most notably: the dissolution of the Soviet Union, with an attendant threat of illicit trade in nuclear materials, technology, and expertise; the discovery following the Gulf War of the magnitude and advanced character of the Iraqi build-up of unconventional weapons;[2] programmes to acquire WMD in certain developing countries; widespread proliferation of ballistic and cruise missiles; and, as yet unanswered questions about the extent and nature of the North Korean nuclear and ballistic missile programmes.

Among these myriad NACD efforts, multilateral export control arrangements operating by consensus and "national discretion," have come under critical scrutiny in recent years. Despite declarations that their purpose is to provide such benefits as clarity, understanding, and harmony, the regimes have been targeted by those who find them "discriminatory" and by those who find them not discriminatory enough. Shahram Chubin has asserted that the Third World views the "host of regimes to control and restrain suppliers" as "all designed to deny certain technologies to Southern states."[3] As Michael Moodie has noted, "regimes such as the MTCR are seen as efforts to deny to the developing world the advanced technologies needed not only for legitimate commercial purposes but as the foundation for sustained development.[4]

1. In this chapter, 'regimes' refers to multilateral export control arrangements. The arrangements discussed in the chapter include: the Nuclear (or London) Suppliers Group (NSG), the Australia Group (AG), the Missile Technology Control Regime (MTCR), and the Wassenaar Arrangement.
2. See *Multilateral Verification and the Post-Gulf Environment: Learning from the UNSCOM Experience. Symposium Proceedings*, Steven Mataija and J. Marshall Beier, eds. (Toronto: Centre for International and Strategic Studies, 1992).
3. Shahram Chubin, "The South and the New World Order," in *Weapons Proliferation in the 1990s*, Brad Roberts, ed. (Cambridge: MIT Press, 1995), 411–432.
4. Michael Moodie, "Beyond Proliferation: The Challenge of Technology Diffusion," in

On the other hand, David Kay, drawing on the case of Iraq, has claimed that "nuclear export controls have proved to be vulnerable to deception."[5] Others have suggested that these regimes need strengthening, formalisation, and/or harmonisation.

While open to signature on a global basis, the major arms control treaties which seek to constrain or prevent proliferation of WMD have not, of course, been immune to complaints. The continuation of the fundamental inequity associated with the Treaty on the Non-Proliferation of Nuclear Weapons (NPT) — the distinction it perpetuates between the nuclear weapons states (NWS) and the non-nuclear weapons states (NNWS)[6] — was a reason for diplomatic sparring during the Conference leading to the permanent extension of the Treaty. While 1995 has come and gone, the "93+2" IAEA Safeguards measures included under Part II of the programme — more particularly, the measures which provide information about, and access to, areas where no nuclear material is located — have thus far been blocked from inclusion in a required Additional Protocol by certain Member States. Political debates over the verifiability of the Chemical Weapons Convention (CWC) and concerns over the costs associated with its verification regime and the required elimination of chemical weapon stockpiles have led to a lengthy period of deliberation regarding its ratification by two key signatories, the United States and the Russian Federation.[7] Negotiation of the Comprehensive Test Ban Treaty (CTBT) was buffeted by concerns which seemed at one point to be irreconcilable within the deadline set by the five declared nuclear powers. Disputes over the scope, nature, and cost-effectiveness of proposed verification and transparency measures intended to strengthen the Biological Weapons Convention (BWC) have delayed completion of a Compliance Protocol. While Member States of the Conference on Disarmament (CD) have agreed on a negotiating mandate for a fissile material cut-off treaty, the demand of a few states in the CD to link cut-off negotiations to other nuclear disarmament issues has brought progress there to a standstill.

In the bilateral area, while the Intermediate-Range Nuclear Forces (INF) Treaty was successfully implemented, and the U.S.-former Soviet Union Cooperative Threat Reduction (CTR) and Material Protection, Control and

---

Weapons Proliferation in the 1990s, 83.

5. David A. Kay, "Denial and Deception Practices of WMD Proliferators: Iraq and Beyond," in *Weapons Proliferation in the 1990s*, 309.
6. These categories of states are often called the nuclear "haves" and "have nots" — terms which emphasize the distinction.
7. The time for political debate in both countries is running out. If the United States and the Russian Federation do not ratify the CWC before it enters into force on 29 April 1997, they will not be original parties to the Convention, will have no voice in deciding the final scope of the treaty, and will not be represented on its executive decision-making body. The Russian Duma has passed a law which provides a legal framework for the destruction of chemical weapons on Russian soil.

Accounting (MPC&A) programmes are progressing, the Duma's ratification of START II has been stalled by Russian opposition to NATO expansion, and negotiations over modifications to the ABM Treaty proposed by the United States have been protracted and heated. Russian concern over NATO expansion has also threatened the updating of the CFE Treaty, and it has been linked to the delay in Russia's ratification of the Open Skies agreement.[8]

Developing Criteria for Evaluating NACD Efforts

Given the number and significance of the NACD agreements, measures, and regimes presently being negotiated and proposed, policy-makers often face a complex task of evaluating their advantages and limitations in order to judge whether signature or membership is in the net best interest of their country. As Andrei Kortunov has pointed out, in an uncertain security environment, it is hard to evaluate new defence and arms control priorities and policies. "Under differing conditions, the same weapons systems might appear stabilizing or destabilizing, the same military programmes economically affordable or not affordable, the same strategies a manifestation of prudence and realism or a display of irresponsibility and recklessness."[9]

The development of evaluative criteria cannot be accomplished in the abstract: the criteria must be placed in context. To give one example, the concept of "military significance" has been used for some years in determining whether it is in the interest of the United States and the former Soviet Union to ratify bilateral agreements. However, military significance is situation-dependent: in the U.S.-Russian context, where there are thousands of weapons of mass destruction, a militarily significant violation might mean a few hundred weapons; in a regional context where there are only a few such weapons, it would be a much lower number. In the former case, it would take a significant shift in the military balance for the bilateral relationship to become destabilizing; in the latter case, a minor change in the number of weapons could become militarily significant.

In evaluating NACD efforts, policy-makers must identify, define, and prioritise national interests. Then they need to apply criteria based on these interests to the specific case.[10] Of primary importance are the country's

---

8. At the U.S.-Russia Summit in Helsinki, many of these issues were apparently resolved. President Clinton announced that NATO enlargement will take place; a NATO-Russia document providing for a forum for regular consultations will be negotiated and signed; the CFE Treaty will be adapted to the post-Cold War era by setting new limits on conventional forces; START III guidelines have been agreed upon; and, a demarcation agreement distinguishing between ABM Treaty-restricted ballistic missile systems and theatre missile defences has been reached. Report of the Joint News Conference of President William J. Clinton and President Boris Yeltsin, 21 March 1997.

9. Andrei Kortunov, "The U.S. and Russia: A Virtual Partnership," *Comparative Strategy* 15:4 (October–December 1996), 339.

national and international security policies. In the Cold War period, the United States and the Soviet Union considered these distinct entities. U.S.-Soviet agreements were negotiated, signed, and ratified because the two parties, despite their adversarial relationship, believed that the agreements enhanced their relative security positions. Such decisions were based in large part upon perceptions of weapons and weapon systems parity between the two parties. Although international approval was desirable, it was certainly less of a consideration than national security. Thus, the ABM Treaty was — and still is — considered to be in the national, economic, and military security interests of its signatories, and the linkage between strategic offensive and strategic defensive arms which the Treaty reflects has been incorporated, some twenty years later, into the START I and START II agreements.

Looking ahead to the 21st century, with emphasis increasingly placed on regional and global agreements and measures, the distinctions between national and multilateral security policies may tend to blur. Indeed, there may be occasions when, in assessing or negotiating an agreement, it becomes clear that the security benefits and costs for the international community may be different from, and even weigh more heavily, than a country's national security interests. In such cases, when the benefits are considered separately, the benefits to the international community may be persuasive enough to "tilt the balance" in favor of an agreement. It might be argued that certain members of the P-5 took into consideration the concerns of the international community in agreeing to modifications in the CTBT drafts which were not supported by their defence and military personnel. On the other hand, the conflict between national and international security goals more often leads to an attempt to block adoption of an agreement; such was the case with India's efforts to prevent the CTBT from becoming law.[11]

Entry into an NACD effort requires weighing the effort in terms of its consistency with the country's near- and long-term foreign policy objectives, including the country's current and desired relations with other countries, with regional organisations, and with international bodies. Possible reactions by the public and by other countries, including allies, need to be analyzed. Basic inconsistencies among the country's foreign policy objectives may need to be identified, for example the conflict between the objective of non-proliferation and the economic gains associated with the sale of arms. Several recent examples might be cited. Sales of Chinese equipment and expertise to Paki-

---

10. Some key evaluative criteria are examined in this chapter. For a checklist of criteria and an expanded discussion of the topic, see *Criteria for Evaluating Potential Agreements or Measures to Control Arms and Promote Stability*, Department of Foreign Affairs and International Trade, Canada, 1997. The report was written by Patricia McFate and Sidney Graybeal.

11. On 23 January 1997, Indian External Affairs Minister Inder Kumar Gujral reiterated that New Delhi would not sign the CTBT because "it was no longer linked to the process of nuclear disarmament."

stan and Iran may be in violation of a promise China made to abide by the terms of the MTCR even though it is not a full member. Syria's effort to buy high-technology upgrades for its T-72 tanks has created a test of the strength of the Wassenaar Arrangement in which the economic and domestic interests of certain forum members conflict with the Arrangement's efforts to regulate arms transfers that might upset regional stability. Russian plans to sell two nuclear reactors to India have prompted assertions that the sale would violate a 1992 "suppliers agreement" of the Nuclear Suppliers Group and would harm global non-proliferation efforts;[12] its sale of SS-4 medium-range missile technology to Iran, if confirmed, would violate the MTCR.[13]

While regional and global security are enhanced by multilateral NACD agreements and regimes, consensus is not without its costs at the *treaty* negotiating table. Membership in a *regime* may not have the same costs; however, consensus at the point of judgment regarding an action may have a negative impact upon a Member State's national or economic security. In such cases, evaluation might focus on the question of whether the regime's framework reflects an appropriate international norm of behaviour and whether its decisions strengthen regional and international peace and security. In addition, there is the question of whether the informal nature of the regime and its lack of enforcement mechanisms could contribute to generating a false sense of security.

World commerce will continue to expand in the 21st century. Thus, the short- and long-term domestic economic benefits and costs of entry into an agreement or a regime must be weighed. Will it result in savings which can be used for domestic priorities? Will it impact bilateral, regional, or international trade agreements already in place? Will it result in unwarranted or heavy burdens upon private companies, such as loss of propriety information, invasion of privacy, or financial costs? The Chemical Weapons Convention is an example of an NACD agreement in which hard questions have been asked about its economic impact not only on the private industry of those countries which sign the agreement, but upon the industries of those countries which do not choose to become Parties to the Convention.

Another aspect of the question of economic costs has been raised by the Russian Federation in facing the costs associated with the dismantlement and

---

12. The 1993 agreement imposed an embargo on sales of nuclear technology to India, Pakistan, and Israel, all of which have refused to sign the NPT and accept international monitoring. The United States and Russia are already at odds over Moscow's decision to sell a nuclear reactor to Iran and Russia's desire to help Cuba complete construction of a nuclear power plant. Information about these events was summarized by Michael R. Gordon in "Russia Selling Atomic Plants to India; U.S. Protests Deal," *New York Times*, 6 February 1997.

13. The sale was reported in the *Washington Times*, 13 February 1997. U.S. Vice President Al Gore raised the issue during talks with Russian Prime Minister Viktor Chernomyrdin after the CIA reported the sale.

elimination of the extensive stockpile of chemical weapons it inherited upon the dissolution of the Soviet Union, and the concerns regarding public safety which will impact the methods and pace of the process. In recognition of the costs which Russia and other republics of the former Soviet Union will incur in fulfilling this and all their other NACD obligations to reduce and eliminate WMD, the U.S. Congress-mandated CTR programme has provided services, tools, and technology.

While the CTR programme is a cooperative bilateral effort based on economic realities, economic costs and benefits can also be part of "coercive diplomacy." As explained by former Secretary of Defense William Perry, in the efforts by the United States to stop the North Korean nuclear weapons programme, Pyongyang was presented with the alternative of economic sanctions if it did not stop its programme and economic assistance in the production of commercial power if it did.[14]

Weighing security and economic benefits is not without its complexities. Weighing security, economic, and humanitarian benefits is even more complicated. Such has been the case in the negotiations associated with the Convention on the Use of Certain Conventional Weapons (CCW). Parties have yet to reach agreement on how to achieve a total global ban of anti-personnel landmines because of the reluctance of certain states to forego sale of landmines, the arguments made by several parties on behalf of the mines' ability to protect their soldiers, and the international commitments and responsibilities cited by one party.[15] While 155 countries voted recently for a non-binding UN General Assembly (UNGA) resolution calling on Member States to complete negotiations on a legally binding, international agreement banning the production, stockpiling, use, and transfer of anti-personnel landmines, ten countries abstained, including China, Pakistan, and Russia, the principal opponents of a global ban.

Military security, including the military security of allies and potential adversaries, must be considered in evaluating an NACD effort. Inevitably, this consideration must take into account the industrial base which supports the country's military capability and the likely reaction of defence and military personnel. In looking at the evaluative criteria of military security, the current efforts to adapt the CFE to the new security environment are of particular

14. Remarks prepared for delivery at the John F. Kennedy School of Government, Harvard University, 13 May 1996.

15. China and Russia already have indicated that they would not sign a global treaty banning the use, export, production, and stockpiling of anti-personnel landmines; they probably will agree to take part in the CD negotiations on the subject, however. President Clinton has named a global ban on anti-personnel landmines as "one of my top arms control priorities"; however, he has insisted that U.S. "international commitments and responsibilities" must be taken into account in any negotiations on a total ban. Letter to the U.S. Senate dated 9 January 1997.

importance. As expected, the core issues will include new redistribution mechanisms for treaty-limited equipment in light of NATO's anticipated expansion to include former Warsaw Pact countries, new provisions on stationing forces, and ways to assure the function of the treaty in cases of crisis and conflict. In the war of words over NATO expansion, the Clinton Administration has countered the perception of NATO as a "military alliance" and "defence treaty" by pointing to its "political function," that is, its role in helping Central European countries to evolve toward democratic societies, market economies, and peaceful neighbours. At the same time, legitimate Russian concerns will of necessity modify decisions at the Joint Consultative Group table when the CFE "adaptation" process is underway.

Significant NACD agreements, measures, and regimes reflect how their signatories wish to respond to threats to national and international security. While the agreement or regime may not eliminate the threat, it should certainly constrain or reduce it. In the area of peace operations, however, conflict management may be intertwined with humanitarian concerns. The threat may not endanger the national security of the countries participating in the operation; rather, the motive for participation may be the wish to reduce the suffering of innocent civilians, while, at the same time, not excessively endangering civilian and military personnel involved in the operation.

In the 21st century, long-held NACD tenets should be the subject of reconsideration. Cited here are two examples.

1) Is control of arms always a "good"? In regional "hot spots," for example, South Asia or the Middle East, could the existence of, or capability to acquire, weapons be a stabilizing deterrent factor which should be considered in regard to any regional NACD agreement?

2) Is enhancing transparency always a desirable outcome? Does it always contribute to increased stability, or could it generate instabilities by revealing military capabilities and limitations which may lead to an arms race or actual hostilities? If a data exchange is included in the agreement, could it have a detrimental effect upon the diplomatic and military relations among the parties to the agreement?

Verifying NACD Efforts

In no NACD area has there been as significant a shift from "old thinking" to "new thinking" than in the field of verification. The advent of national technical means (NTM) led to the pragmatic approach toward monitoring taken by the United States and the Soviet Union during the 1960s and 1970s. Support for highly extensive and intrusive verification means, thought by some major powers to be a necessity in the 1980s, has been tempered by the recognition of the costs of such verification, including security, financial, human resources, and technology costs. In the 21st century, additional transparency measures, including voluntary exchanges of information, invitational access to sites and facilities, and cooperative implementation of agreed steps to cut back existing military postures, should significantly increase openness and build confidence.

Cooperative monitoring — the collection, analysis, and sharing of information among parties to an agreement, measure, or regime — will be a viable option in many new regional and global contexts if the technologies incorporated into the monitoring regime are shared among all parties, and all parties receive equal access to data acquired by the system. Cooperative monitoring technologies facilitate implementation of new NACD cooperative security efforts by providing the capability to observe relevant activities, to define and measure agreed-upon parameters, to record and manage information, and to carry out inspections using agreed-upon standardised monitoring systems. Before cooperative monitoring measures are put into place, joint experimental approaches to develop cooperative technology may be required to encourage such sharing; these approaches cannot only preceed formal agreements, but, as Arian Pregenzer has pointed out, they can provide information needed by negotiators during the negotiating process.[16] In addition to support of NACD efforts, cooperative monitoring systems can be applied to other areas, for example, the civilian nuclear fuel cycle, borders and demilitarized zones, and acid rain and water pollution. Already regional monitoring of radiation levels in the Sea of Japan is taking place among the IAEA, the Republic of Korea, Japan, and Russia. China and the ROK are monitoring air quality over the Yellow Sea, and the ROK, China, and Japan are sharing environmental radiation data.

One determining factor in answering the question of how much verification is enough is the context in which the regime must operate. If verification means in the 21st century reflect the environment in which they are taking place, they will be more cost-effective. In a cooperative environment, verification measures should be designed to confirm compliance and resolve ambiguities; in an adversarial or coercive environment, verification's purposes have been deterrence and hopefully *early* detection of non-compliance. If relations between the parties deteriorate, the agreement should be robust enough to permit the implementing body to investigate ambiguous activities and anomalous events.

The degree of intrusiveness and the possibility of cooperative measures should depend upon the situation. A graduated system of verification actions that would allow for less or more extensive and intrusive verification measures and techniques depending upon *agreed* indications of compliance or non-compliance would be more cost-effective than the costly and burdensome verification regimes of the Cold War and post-Cold War.[17]

---

16. Arian L. Pregenzer, "Cooperative Security: The Role of Technology," Sandia National Laboratories, undated paper.
17. Such a graduated system of verification actions, based on the U.S. national security system used to describe defence condition (DEFCON) was proposed by the author in

In the Cold War and post-Cold War periods, relative equalities in the monitoring capabilities of the United States and the former Soviet Union permitted negotiation of bilateral agreements. In the negotiation of regional and global NACD agreements, however, disparities in national means have become a focus of debate and have led to contention. During the CTBT negotiations, for example, there were significant differences over verification, primarily on the subjects of the degree of intrusion allowable, the procedures for on-site inspections (OSIs), and whether the use of information derived from NTM should be used as a basis on which to demand inspections.

Even in situations in which parties with advanced capabilities may be willing to share some collection products, other parties must decide whether to accept the products without the ability to understand and participate in the collection, the data reduction, and the processing techniques, or to independently analyse the raw data. The question comes down to this: if it is necessary to degrade the product to protect "sources and methods," will other parties be willing to accept the degraded product? Recent advances in the concept of cooperative monitoring will certainly play a positive role in resolving some of these issues in the future.

Countries entering into agreements, measures, or regimes in the 21st century should be prepared to perform cost-benefit analyses in which the benefits of the effort, such as enhanced transparency, and national, regional, and/or international peace and security, should be weighed against the costs of implementing the agreement — for example, loss of sensitive data, personnel, equipment, and organisational costs, and loss of proprietary information. Analysts should factor in the synergistic benefits of multiple methods of verification within the NACD effort and across other efforts. Methods and measures used to verify the agreement may be mutually reinforcing. The verification efforts of other regimes may be strengthened by this effort's verification regime. There also may be synergistic effects associated with verifying the agreement that could be used to enhance the viability and utility of other NACD efforts and of regional and international organisations.

The Roles of Outer Space and Cyberspace

The impact of outer space and cyberspace on NACD efforts in the 21st century will be profound; indeed, the NACD 'Information Age' has already begun. Shared information technologies have been of particular relevance in the arena of UN peace operations. Overhead imagery has been used by UNSCOM and IAEA inspection teams in Iraq and by the IAEA with regard to the North Korean case as an important supplement to other verification activities; the imagery has assisted these international bodies in reaching independent deci-

---

"Where Do We Go from Here? Verifying Future Arms Control Agreements," *The Washington Quarterly* 15:4 (Autumn 1992), 84.

sions concerning verification matters. The technologies involved in the Iraq case, including satellite imagery from both NTM and commercial sources, high altitude aerial imagery, medium altitude aerial imagery, and imagery from helicopters, could well be of utility in future monitoring regimes involving the United Nations.[18] Military and civilian personnel involved in Bosnia pursuint to the Dayton Accords have also benefited from shared intelligence collection, surveillance, and reconnaissance.

Since April 1996, the United States has provided NATO allies access to warnings from U.S. sentry satellites of Scud-type missile attacks. The information includes electronic intelligence, imagery, and access to a set of databases containing information such as orders of battle and situational reporting. In the near future, the Russian-American Observation Satellite (RAMOS) will provide satellite data associated with the launch of ballistic missiles. RAMOS could play an important role in the future by providing strategic warning during periods of regional conflict. By the 21st century, if Russia and the United States were to combine elements of their early warning systems, they could significantly improve their ability to detect and track missile launches and missiles and warheads in flight over key regions associated with the spread and potential use of missile technologies for delivery of WMD. In addition, U.S.-Russian cooperative space-based intelligence gathering could increase the overall effectiveness of technical means dedicated to monitoring proliferation and other forms of disturbing or hostile regional military activity. These cooperative activities will depend, of course, upon achieving high levels of mutual confidence between two former adversaries.

The threats to international security and stability in the 21st century cannot all be resolved by NACD efforts alone. Crime syndicates will flourish; drug trafficking, money laundering, financial fraud, and corruption will increase; terrorists will roam freely via global networks; and, epidemic diseases will cross state borders. Global intelligence and information links will be needed to combat these threats. These links may have a beneficial synergistic effect when combined with NACD verification measures and systems.

Threats to the international community will be a key concern of the United Nations at the beginning of its second half century. With its nearly global membership, the United Nations remains the most significant international body focussed on global peace and security. In a recent study, four Canadian and American authors proposed the development of a Centre for Information, Training, and Analysis (CITA) that could serve as a resource for the Secretary General in his increasing responsibilities in the areas of arms control, disarmament, confidence-building, and conflict management.[19] CITA would

18. The UNSCOM and IAEA monitoring operations in Iraq could be called a "verification laboratory."

19. Patricia Bliss McFate, F. Ronald Cleminson, Sidney N. Graybeal, and George R. Lindsey, *Verification in a Global Context: The Establishment and Operation of a*

acquire and collate information of importance to the Secretary General in these areas and it would provide analyses which relate to verification of compliance with obligations for which he is responsible. In addition, it would provide information on the field of verification, including instruction in the reduction, interpretation, and analysis of information from a variety of sensors; cooperative monitoring technologies; the conduct of verification measures; and data management information systems. As conceived, CITA would be a lean, mission-oriented unit which could capitalise upon the ability of the United Nations to draw upon a wide range of sources for verification-related information. In a few years, if the new Secretary General tightens the UN budget, restructures the organisation, tames its bureaucratic struggles, and effectively articulates its mission to Member States, he should make the case for a unit which performs the functions envisioned for CITA.

New technologies based in cyberspace will also enhance the ability to verify the provisions of NACD agreements, measures, or regimes. Research in increasingly sophisticated micro-electro-mechanical systems (MEMS), for example, has already led to the development of inexpensive chemical analysis instruments — the equivalent of a chemical laboratory on a chip. This research will lead, in turn, to inexpensive, portable devices for detecting chemical and biological weapons that can be carried by soldiers or by on-site inspectors. Another example would be "Straight-Line," a pilot system of sensors being developed by Sandia National Laboratories which gathers information from stored nuclear material and distributes both classified and unclassified data to authorized users on a strict need-to-know basis.

Advanced computer systems have permitted the development of data management systems to handle the information which will be exchanged in complex NACD agreements, measures, or regimes. In addition, the computerised systems which have provided training for soldiers on "virtual battlefields" can also provide training through the use of "virtual inspections."

The technologies of the Information Age hold both promise and danger. While advancing information and communication technologies may provide the means to minimise or negate the threat, they may also increase the threat — or indeed be the threat. NAVSTAR GPS and GLONASS signals can provide high accuracy navigation for civilian or military purposes. Proponents of commercial use of former military spy satellites envision using the images for vacation planning or business-related demographics; opponents fear that Libya, North Korea, and other countries without reconnaissance systems will use the images for military purposes. Environmental monitoring satellites can provide the data to report the sources of environmental degradation which lead to regional conflicts, or they can serve as force-multipliers when used for military purposes. The hardware, technology, and production facilities for

---

United Nations Centre for Information, Training, and Analysis (CITA), Arms Control Verification Studies, no. 7 (Ottawa: Department of Foreign Affairs, 1996).

space launch vehicles are essentially interchangeable with those for ballistic missiles.[20]

The spread of the Internet, low cost computing power, and the global mobility of technologically sophisticated individuals have all functioned to promote knowledge, but they also have shrunk the distance a proliferant state must travel to acquire the necessary knowledge to construct missiles and WMD.[21] Information technology can undermine repressive governments. However, digital terrorism is a threat to information networks. In the 21st century, on both the national and multilateral levels, infrastructure dependencies and vulnerabilities will have to be addressed, threat conditions defined, response options explored, and information systems protected.

Unquestionably, advancements in verification technology will lead to cost-effective measures, but they will also lead to more compliance questions. In treaties, mechanisms are usually provided for resolution of compliance. Implementing bodies determine whether the perceived non-compliance is a mistake in interpretation, a technical error, or an act of deliberate non-compliance. In the case of agreements in which actions — remedial or punitive — in response to non-compliance have not been spelled out in the text of the treaty, recourse to diplomatic, economic, or military action on the part of the other signatories is an option. The ultimate recourse may be an appeal to the UN Security Council.

In the export regimes, there are no agreed-upon verification measures. Resolution of compliance concerns is a private matter among the parties to the regime. Often interpretational disputes can arise. The MTCR, for example, has been an effective instrument in constraining certain cases of proliferation by causing missile programme dismantlement, cancellation, conversion, and schedule delays, as well as increasing the "costs" — in all senses of the word — of missile acquisition. Enforcement of the regime's provisions has not only slowed the emergence of new states acquiring ballistic missiles, but the list of suppliers has also shrunk.[22] However significant these accomplishments, the MTCR remains a voluntary cartel, with guidelines which are not enforceable. Final decisions on exports ultimately become domestic decisions in which politi-

---

20. The MTCR Guidelines permit its members to support the space programmes of other countries or international cooperation in such a programme "as long as such programmes could not contribute to delivery systems for weapons of mass destruction."
21. Former CIA Director John Deutsch has called computer warfare the second most worrisome threat to national security, after nuclear or chemical attack. The U.S. Department of Defense's Defense Science Board has predicted that by 2005, information-related attacks by terrorists, organised criminals, and foreign spies will be "widespread." "Panel of Experts Warn of the Threat of Computer Warfare," *Chicago Tribune*, 8 February 1997.
22. North Korea has become the main supplier of MTCR-restricted ballistic missiles. China also apparently continues to supply MTCR-restricted ballistic missiles, as well as missile technology and equipment to a limited number of customers, including Pakistan and Iran.

cal and economic considerations are weighed along with national and international security concerns. Obligations are not verified by the Regime, nor is the end-use of the recipient states. "Adherence" to the guidelines varies widely and interpretational disputes arise, given the vague nature of their wording.

Can NACD Efforts Be Harmonised?

As the next century approaches, NACD efforts will continue to multiply in response to particular concerns, such as countries which seek to acquire more than one type of WMD and advanced delivery means and rogue states with rapidly mutating weapons programmes, moving from less sophisticated to more sophisticated weaponry as they acquire materials and expertise. Additive budgets and bureaucratic proliferation argue for more cost-effective NACD implementation, but it may require breaking the mold of a separate implementation body for every treaty. One possible step would be to designate a strengthened IAEA as the implementing body for a future fissile materials cut-off treaty. Another possibility would be to link the Organisation for the Prohibition of Chemical Weapons (OPCW) with a future body responsible for implementing the BWC verification provisions. Better coordination between implementing bodies could save costs and be mutually beneficial, although there would be problems in achieving effective coordination.

Surely the benefits of shared intelligence, shared enforcement, and shared reinforcement of global norms of non-proliferation argue for such harmonisation. On the other hand, not all countries belong to all the agreements and regimes, the bureaucracy associated with harmonisation might be extremely costly, the lack of consensus among countries regarding policy objectives and economic gains associated with arms sales and transfers might bog down the enterprise, and then there is the "special case" of the NPT, with its division between nuclear weapons states and non-nuclear weapons states which bears no correspondence to the CWC or BWC.

Proposals for reforming export regimes have been put forward by a number of authors. Some have proposed fusion, others have suggested the importance of integrating and simplifying the existing export controls. In their 1996 report, Spector and Foran examined three approaches for the possible integration of the export control regimes: (1) unification into a single comprehensive export control monitoring scheme; (2) keeping the regimes separate while incorporating the most effective practices of each into all; and, (3) keeping them separate while harmonising to ensure consistency of approach and avoid undercutting.[23] Spector and Foran concluded the first approach would not work, but a combination of the last two options was possible.

---

23. Leonard Spector and Virginia Foran, *Preventing Weapons Proliferation: Should the Regimes Be Combined?* A report prepared for The Thirty-Third Strategy for Peace, U.S. Foreign Policy Conference, Warrenton, Virgina, 22–24 October 1996. Other studies are cited by Gary K. Bertsch and Richard T. Cupitt, "Non Proliferation in the

Gary Bertsch and Richard Cupitt have looked at patterns of membership and adherence to four export regimes — COCOM, the NSG, the MTCR, and the AG — and divided countries into five groups: coordinating, collaborating, sensitive, threatening, and peripheral states.[24] In this hierarchy, coordinating states are those which belong to all four regimes. They then propose six sometimes complementary strategies available for the export control policies of the coordinating states: harmonisation, liberalisation, safeguards, assistance, conditional aid, and denial.

Some have proposed the harmonisation of a regime and treaty in the same weapons area, but this would only underline the gulf of differences between treaties and regimes. Take, for example, the AG and the CWC. The AG is an informal forum of states whose goal is to discourage and impede chemical weapons proliferation by harmonising national export controls on CW precursor chemicals, sharing information on target countries, and seeking other ways to curb the use of CW. The CWC is a legally-binding global treaty that bans the production, acquisition, stockpiling, and use of chemical weapons. The Australia Group has no charter or constitution, no verification regime, and no implementing body; it operates by consensus. The CWC has an extensive and rigorous verification regime, an explicit and complex implementation body, and enforcement mechanisms, including penalties for not joining the Convention, are written into its text. The two efforts cannot be harmonised without turning the AG into an adjunct to the CWC, a move which would require a lengthy negotiation process and enormous bureaucratic resources.

While some members of the Non-aligned Movement, especially Iran, Pakistan, and Indonesia, have argued that the NSG and AG are inconsistent with their respective treaties, the NPT and CWC, these regimes should not be dismissed: they set and reinforce norms of global behaviour embodied in legally-binding agreements; politically-binding export regimes provide flexible structures in which the differences among countries can be accepted.

The MTCR and the Wassenaar Arrangement operate in areas where treaties do not exist. While there is widespread recognition of the role which ballistic and cruise missiles may play in the delivery of WMD, there is little international agreement on whether or how to limit acquisition of these delivery systems. There is even less agreement on how to limit the sale of advanced conventional weapons — indeed, there is little agreement on what to limit. Some policy-makers have proposed limits on weapons with certain characteristics, including autonomous — fire-and-forget — operation, high accuracy, long-range, and/or the ability to defeat countermeasures.[25] Others have tar-

---

1990s: Enhancing International Cooperation on Export Controls," in *Weapons Proliferation in the 1990s*, 119–136.

24. Bertsch and Cupitt, *op. cit*, 125–126.

25. Examples would include stealth aircraft, advanced ballistic and cruise missiles, and directed energy weapons.

geted "weapons of ill-repute," such as incendiary and fragmentation weapons, blinding lasers, and anti-personnel landmines.

The MTCR and the Wassenaar Arrangement function in a diffuse and changing environment in which the sources of advanced conventional weapons and dual-use technologies are steadily expanding, critical technologies vital to defence are increasingly commercial in origin, and economic stresses have led to aggressive selling by some countries of products once reserved for their militaries. Proposals to turn these suppliers regimes into global treaties appear unlikely to succeed in the near future because the members of the regimes do not evaluate national, regional, and international security threats in the same way. Efforts by one country to block the sale of technology often backfire when other countries do not share the same point of view. An example of this is the recent purchase of an IBM supercomputer by Russia's Ministry of Atomic Energy Affairs. When the United States blocked the sale of the computer by the American computer manufacturer, it was obtained through a European intermediary.

Enhancing NACD Efforts: Four Proposals

Given the challenges for verification and implementation of treaties and regimes, how might these NACD efforts be enhanced in the 21st century? Here are four 'modest' proposals.

First, expand the review processes of the export regimes to permit a more extensive dialogue between sellers and buyers in which the reasons for purchasing materials, technologies, equipment, or systems could be thoroughly explored. In this way, the emphasis would move from "discriminatory" or "arbitrary" restraints imposed by supplier cartels to discussion of non-military means of resolving conflicts, and from controls on exports to controls on their application or end-use. Evaluative criteria for approving export might include the following: the contribution which the sale makes to regional stability; the buyer's participation in relevant multilateral NACD efforts; the lack of evidence of clandestine or illegal purchases of prohibited items, or abuse of authorised purchases; the lack of a previous record of being denied purchase of authorised materials, equipment, or technologies; and, the purpose for which the materials, items, or technologies will be used.

Second, perform experiments in cooperative monitoring of dual-use exports. Interrogative tags, or other forms of identification, might be placed on the items or equipment, so that the end-use of the export could be monitored. This proposal would offer advantages both to the shipper and to the recipient: the shipper could determine that the end-use complied with the restrictions of the export regime, and the recipient, by demonstrating compliance, could assure the international community of its allegiance to non-proliferation goals.

Third, develop a common, global system of accounting and physical security standards in order to monitor and track the birth-to-death cycle of direct-use nuclear materials. The system might have its genesis in the MPC&A cooperative programmes now in operation or under development between the United States and the Russian Federation. Expanding this system to include all relevant countries would contribute to international stability and security.

Fourth, build upon the recognised relationship between regional security and economic growth. Recently, the International Bank for Reconstruction and Development and the International Monetary Fund have linked international financial assistance with norms of military behaviour, including defence expenditures and compliance with treaties. This concept has already been put to a test. Yaron Ezrahi noted recently that news of the signature of the Hebron agreement was overshadowed in Israel by the IMF's announcement that it will reclassify Israel as an economically advanced state. As he pointed out, "both the Hebron agreement and the IMF decision reflect Israel's readiness to choose the economy over ideology. ... For many Israelis, the [Hebron] agreement was also a moral choice between the nonviolent challenges of the global marketplace and the bloody challenges of the battlefield."[26] The IMF might be encouraged to expand the direction of its initiative by offering to invest in the operational safety of facilities which contain dual-use materials if a country complies with NACD goals and norms.

Conclusion

Existing NACD efforts have had different histories and different mechanisms for control, enforcement, and administration. Despite their separate existence, they face the same challenge: proscribed weapons, weapon delivery systems, and dual-use technologies are proliferating, often involving the same arms exporters and the same recipients. Periodic consultations, shared intelligence, a concerted effort to marshal support from the international community for enforcement of global norms of non-proliferation, and, where appropriate, more inclusive consultation with potential recipients would be modest ways to harmonise and address the concern of another Iraq-type situation. The synergistic effects will be higher than the bureaucratic costs.

By the 21st century, the verification and implementing bodies and forums associated with NACD efforts will be an enormous burden — in terms of costs, personnel, and equipment — to the countries which participate in these efforts. On the other hand, until the significant number of countries which seek or have acquired WMD and which acquire or sell advanced delivery systems give up these pursuits, it may continue to be a necessary burden.[27]

---

26. "In Israel, Peace Means Prosperity," *The New York Times*, 21 January 1997.

27. At the present time, approximately twenty-four countries are "actively engaged in the development of nuclear, chemical, or biological weapons." Lt.Gen. Patrick Hughes, USA, Director Defense Intelligence Agency, *Global Military Threats to the United*

Participation in NACD efforts, including agreements, measures, and regimes has been relatively low cost if they are compared to expensive and risky military measures. Over time, these expenditures can be reduced as the norms of non-proliferation, arms control, and disarmament become the global standard of conduct.

States and Its Interests Abroad, statement for the Senate Select Committee on Intelligence, 22 February 1996, 22.

Chapter Four

Significant Multilateral NACD Agreements: The Scope and Challenge of Implementation

Richard Guthrie*

While recognising that it is impossible to cover everything, the following is an attempt to present an overview of what is happening in the field of non-proliferation, arms control, and disarmament, with a particular focus on what may be the significant issues of the next few years. This chapter shall first discuss regional treaty regimes then move on to global treaty regimes.

Regional Arrangements

The Conventional Armed Forces in Europe (CFE) Treaty

The CFE is a paradoxical treaty. Its successes have led to its greatest weaknesses and its weaknesses have led to its greatest successes. Its success in limiting the forces in certain areas in what used to be the Soviet Union has led to tremendous pressures for its renegotiation. Its weakness in being unable to follow true bloc-to-bloc verification arrangements has assisted in the confident dissolution of the former Warsaw Treaty Organisation.

The CFE was negotiated at the thaw of the Cold War. Its negotiation was made possible by the very forces that were soon to make it somewhat outdated. But the CFE has not yet outlived its usefulness. Some may say that it adds little to current stability that is not on offer elsewhere — but this is due to the high level of military stability in Europe at the present in which no state is in serious fear of invasion by a neighbour for the foreseeable future. However, the future is not necessarily so rosy. It is at times when a situation has potential to deteriorate that measures such as the CFE have, perhaps, their greatest value. The capability to resolve accidents and misunderstandings in communications between states and to rein in ambitions of leaders that may have destabilising objectives may keep a lid on what could, otherwise, be a disastrous situation.

---

* This chapter is an edited version of a presentation made to the Fourteenth Annual Ottawa NACD Verification Symposium, "Cyberspace and Outer Space: Transitional Challenges for Multilateral Verification in the 21st Century," 12–15 March 1997, Montebello, Quebec, Canada.

The Future of the CFE Treaty. The desire of the Russian Federation to see aspects of the Treaty opened for renegotiation is being fulfilled. But these renegotiations are sure to be overtaken by other issues.[1] A central question that has to be pursued is that of whether it would be better to amend the existing treaty or negotiate a new one. The advantage of amendment is that the basic treaty structure may be retained and it is a known quantity; the advantages of a new treaty include being able to start afresh and being able to have a clearer entry-into-force arrangement for the new measures, rather than having to pursue potentially messy amendment procedures. In addition, if negotiations were to stall on a new treaty, it would be possible to move to the amendment option, but it is not so easy to move in the opposite direction.

The proposal, in February 1997, from the NATO states for the new negotiations calls for the bloc system to be dropped entirely and for its replacement by a system based on national equipment ceilings. This would be supplemented by a system of territorial ceilings to include equipment deployed from other states — e.g., U.S. forces in Germany — and by the flank zone obligations. On one level, these seem eminently sensible; the bloc system is clearly outdated. However, on another, the reliance on national limits could cause problems following NATO expansion. Two issues feature strongly: the addition of expansion-state forces into the NATO Integrated Military Structure (IMS) and cascading.

How can the negotiations cope with a system of national equipment limits, on the one hand, and a system of collective defence on the other? While those of us familiar with the politics of the West can be confident that there is no plan to invade the Russian heartland, can the Russians be confident of this? With additional forces joining the NATO IMS, in two or more waves, the Russian, Belorussian, and Ukrainian perceptions of their own security situation may become destabilising. Add to this the probability that the new NATO states will replace much of their existing equipment with materiel 'cascaded' from existing NATO states, and this means that the non-NATO states will face numerically and technically superior forces.

Verification: Problems and Possibilities. There is an extremely serious point to be made here about the reduced effectiveness of the verification regime —

---

1. If elements of the Treaty text are being renegotiated, perhaps negotiators could start using the more literally-correct acronym of CAFE for the discussions, as this would give acronym writers much-needed opportunities to exercise their art. For example, additional bean counting could be done under the rubric of Conventional Armed Forces in Europe Auxiliary Unit Limits for Applying Integrated Totals (CAFE AU LAIT), new verification measures could be given names such as the Executive System for the Preservation of Regulations for European Sustained Schemes of Obligations (ESPRESSO), or the Combined Application, Promotion, and Preservation of Uniform Compliance and Conformity with International Norms Organisation (CAPPUCCINO). Those hungry for more could support the Broad-based Integrated System for Counting Under International Treaties (BISCUIT).

unless significant changes are negotiated. Almost certainly, the renegotiations will be premised on three factors: each state will have an 'active quota' limiting the number of inspections it may initiate; each state will have a 'passive quota' limiting the number of inspections it may host; and, groups of states will be able to collaborate on joint inspections and the exchange of information gained from verification activities.

The CFE currently has 30 States Parties. A quota system suitable for the CFE with sixteen members of NATO may suffer once this total has been raised to twenty in the first wave of expansion. The addition of a further two NATO CFE States Parties at a later stage would mean that over two-thirds of the CFE states would be in one grouping. As the NATO states would surely continue with CFE verification collaboration, as they have done since entry-into-force of the Treaty, the pressure on the active quotas of the small number of non-NATO states may be such that confidence in the verification system for those states becomes so low as to make it ineffective. Additional complications would arise if Austria, Estonia, Latvia, Lithuania, and Slovenia were to join NATO — each of these states has already expressed an interest. None of these are currently CFE parties, and it seems implausible that NATO membership would not be accompanied by CFE membership. This could lead to a CFE Treaty with 35 States Parties, of which 27 would be NATO members. Almost certainly, new CFE negotiations would have to be carried out at each new intake of NATO members.

Whatever arrangements that may be introduced to allow a certain amount of flexibility in the verification regime to accommodate NATO expansion could lead to greatly increased costs for the few states carrying out sufficient inspections to be confident of CFE compliance in the much larger number and may be prohibitive. The establishment of an international organisation for verification would be possible, but might be expensive. It would have fewer confidentiality concerns than, say, the IAEA or OPCW, as it would not be dealing with matters of commercial confidentiality — the CFE declarations are already public documents. This would simplify the flow of personnel between an organisation and national governments, so inspectors could serve short tours on secondment from national military service — thus reducing costs. Such an organisation could be tasked with carrying out a specified number of inspections in each state — which would relate to the level of forces declared — and reporting whether a Party's activities are in compliance with its declarations. The Verification Co-operation Centre in Brussels is used by NATO and non-NATO states to arrange inspections. It could form the nucleus of a new body, although its removal from NATO headquarters to a more neutral setting would be beneficial.

Aside from CFE structures, there are factors that should allow some CFE verification costs to be substantially reduced. For example, full implementation of the Open Skies Treaty would allow a reduced number of on-site inspections for maintenance of the current level of confidence in the CFE verification regime. Open Skies entry-into-force currently requires ratifications from Russia, Belarus, and Ukraine.

Nuclear-Weapon-Free Zones

There are four Nuclear-Weapon-Free Zones (NWFZ) treaties covering populated areas: Latin America (Tlatelolco); South Pacific (Rarotonga); Southeast Asia (Bangkok); and, Africa (Pelindaba).

There are some important things to note here. Tlatelolco, the first populated NWFZ, signed in 1967, now has all but one of its territorial components as Parties — the last, Cuba, has signed the Treaty but not yet ratified it. The Treaty is in force for all other regional states as they have each declared that they will allow application of obligations before the Treaty itself is fully in force. Rarotonga, signed in 1985, was rapidly ratified by all of the states of the geographical area covered. However, one nuclear weapons state, France, continued testing nuclear weapons within the zone of application until 1996. Bangkok was signed in 1995 after a very rapid set of negotiations and without adequate discussions on aspects such as the protocol to be signed by the nuclear weapons states. This is leading to a period of possible renegotiation. Finally, Pelindaba was signed in 1996 after decades of negotiation. One of the key factors that allowed progress on this zone was the renunciation of nuclear weapons by South Africa on 24 March 1993.

All of these treaties require states parties to have a comprehensive, or full-scope, safeguards agreement with the International Atomic Energy Agency (IAEA). This is equivalent to the obligation on non-nuclear weapons States Parties to the Nuclear Non-Proliferation Treaty (NPT), which shall be dealt with later.

Global Arrangements

The global arrangements to be discussed here all relate to weapons of mass destruction (WMD) — that is, nuclear, chemical, and biological weapons. To be pedantic, there is a fourth class of WMD: radiological weapons, which are not explicitly prohibited by any global treaty.

Nuclear Testing

The year 1996 saw the conclusion of the Comprehensive Test Ban Treaty (CTBT). It was opened for signature on 24 September 1996 following the initiative of Australia to take the text to the United Nations after attempts to achieve consensus in the negotiations at the Conference on Disarmament (CD) in Geneva had failed. It is worth looking at some of the reasons for the lack of consensus as these may have a long-term impact upon verification.

Disarmament versus Non-Proliferation. One key difficulty in the CTBT negotiations was the diversity of views as to whether the test ban was to be a disarmament or non-proliferation measure. There was a clear distinction between the views of the three 'holdout' states — India, Israel, and Pakistan — and the five nuclear weapons states — China, France, Russia, the UK, and the U.S. The holdout states wanted the CTBT to lead to nuclear disarmament, while the nuclear weapons states saw the CTBT as a means to bring the holdout states into the non-proliferation regime. This was the crux of the matter in the conflict over entry-into-force provisions. It is also worth noting that although India may have expressed its objections to the CTBT in disarmament terms, its concerns were not dissimilar to those that might be expected of a minor nuclear power eager to keep its nuclear weapons options open.

Doublethink. Some of the lack of consensus may be put down to 'double-think' — the Orwellian concept whereby people are proud to believe in contradictory ideas. Take the following, paraphrased, statement on why the CTBT is not really worth supporting:

> The CTBT is not worth supporting as its provisions are so lax as to allow a rogue state to develop and maintain nuclear weapons that could threaten the United States, the United Kingdom, and France and this would be a grave threat to national security.

Many people from nuclear weapons states who support that statement have also made the following, paraphrased, statement as to why the CTBT is not really worth supporting:

> The CTBT is not worth supporting as its provisions are so strict as to prevent the United States, the United Kingdom, and France from developing and maintaining nuclear weapons that could threaten and deter a rogue state and this would be a grave threat to national security.

This truly is doublethink as it is hard to conceive of provisions that are both too lax and too strict at the same time and this indicates that prejudice against arms control is still alive and well in the establishments of the nuclear weapons states.

Verification issues. Owing to the disagreements over entry-into-force provisions for the CTBT, it is not clear when, and perhaps if, the Treaty will become operational. This has direct implications for the verification arrangements as the International Monitoring System — another IMS — has significant start-up costs. The international verification organisation, the CTBTO, and its Preparatory Commission, will require ongoing financing. Some states may decide that they would wish to defer some of these costs, thus limiting the capabilities of the IMS to detect nuclear tests before entry-into-force of the Treaty. There has been talk of provisionally applying the CTBT, much as was done with the CFE. This would raise further questions of cost allocations.

Nuclear Non-Proliferation and Disarmament

The cornerstone of the nuclear non-proliferation regime is the 1968 Nuclear Non-Proliferation Treaty which was given an indefinite extension at its

Review and Extension Conference in 1995. The decision on extension was not taken on its own but was part of a package of measures that included a strengthened review process and the Principles and Objectives document. It is important to remember that indefinite extension does not mean a permanent Treaty but means a Treaty of indeterminate duration that will continue in existence until it is replaced or supplanted by a new measure or measures — this may be in 5, 25, or 100 years time, but its time will come. It may be assumed that the new measure or measures would be whatever was legally required to move to a nuclear-weapons-free world.

Safeguards

All non-nuclear weapons States Parties to the NPT, as well as States Parties to NWFZs, are obliged to negotiate full-scope safeguards agreements with the International Atomic Energy Agency. Following lessons learned from the cases of Iraq, the DPRK, and South Africa, the IAEA has embarked on a project to improve safeguards, known as the "93+2" programme. The "93+2" proposals are divided into two classes: Part I and Part II.

Part I measures are those that may be carried out under the existing full-scope safeguards agreements and remain focused on accountancy of nuclear material with measures to economise on personnel time and enhance the detection of unreported activities. There is widespread agreement on Part I measures. Part II measures, however, contain controversial elements. These represent a move away from traditional nuclear material accountancy through the enhanced capability of the Agency to gain a wider view of nuclear-related activities between, as well as within, states and allowing IAEA inspectors greater access to both declared and undeclared facilities. Part II measures also include powers to take samples from the environment in order to detect radioisotopes given off during nuclear activities, thus allowing the Agency to check whether this matches the levels produced by declared activities. The full package of Part II measures has yet to be agreed.

Identification of Holdout States

As the membership of the NPT has approached universality, the states that decide that they do not wish to accept the international norms in this field are effectively forced into making an overt statement to this end. Of the states not party to the NPT at this point, only three — India, Israel, and Pakistan — are not part of any regional nuclear-weapon-free zones. The nuclear weapons holdout states are thus clearly identifiable, and what they are holding out from is the requirement to comply with full-scope safeguards. This has consequences for the pressures that these states will come under to conform to the norm. Take, for example, the situation at the end of the 1995 NPT Conference. A group of Middle Eastern states brought forth a resolution condemning Israel for not being a party to the Treaty. Allies of Israel, most notably the U.S., were able to argue that to isolate one Middle East state not party to the Treaty was wrong — thus, no individual state was mentioned in the resolution as finally passed. Now that all other states in the region are NPT Parties, it will be extremely difficult to argue against naming Israel.

This concept of isolating holdouts will become important once membership of other WMD treaties becomes near-universal. This will allow even greater pressure to be exerted upon them to conform with international norms.

A Nuclear Weapons-Free World

The international norm on nuclear weapons is their elimination. However, a nuclear-weapons-free world will occur only under one set of circumstances — that is when the nuclear weapons states see it as being in their political and security interests to secure such a policy. Verification will play an important role here, but new models and structures will be required.

Planning for a 'nuclear weapons convention' to mirror those on chemical and biological weapons will inevitably lead nowhere as a new global convention that could deal with the security concerns at the transition to a nuclear-weapons-free world, treating every state as equal, is, frankly, unnegotiable.

However, such a convention could be negotiated to operate once all nuclear weapons have been dismantled, but even then much reinventing of the wheel is taking place. For example, there has been discussion of the establishment of a new international organisation when the IAEA already exists — it would be easier to adapt the IAEA than to start afresh, both politically and economically.

All of this is horrendously premature and is missing the key point. Those who favour a nuclear-weapons-free world would be better off spending their time persuading the nuclear weapons states that it is in their political and security interests to disarm. A new model that could be pursued would be to have a five-power treaty to disarm with the state-on-state verification verified by the IAEA — much as has been shakily evolving in the chemical weapons field.

Chemical and Biological Weapons

It is convenient to move on from the concept of a nuclear-weapons-free world to the issues of chemical and biological weapons, not only because of new models but also as it may be argued that a nuclear-weapons-free world is unachievable while these other weapons remain as a threat. While the CBW threat argument is often used as a smokescreen against a nuclear-weapons-free world, there are some honestly-held fears about this — fears that have to be addressed.

Before looking at each of the CBW regimes, it is worth enquiring into the memberships of the WMD treaties as a whole. In addition to the NPT, there are three further treaties at issue here: the 1925 Geneva Protocol; the 1972 Biological and Toxin Weapons Convention (BTWC); and, the 1993 Chemical Weapons Convention (CWC).

It has already been pointed out that the NPT is starting to approach universality, but how many states are parties to all four WMD treaties? As of 31 December 1996, of 194 states in the world, only 45 (23 per cent) are parties to all four treaties. A further 49 states (25 per cent) have signed, but not yet ratified the CWC and are parties to the other three treaties. This leaves exactly 100 states (51 per cent) remaining that are not signatories or parties to all of the four WMD treaties. It is clear there is a long way to go here.

The Geneva Protocol

The Geneva Protocol prohibits the use, but not the possession, of chemical and biological weapons. Its breadth is considerable because of the simple language used in the text. Three things are of note here. First, it has no compliance mechanisms. Second, many Parties entered reservations to the Protocol on the use of CBW in retaliation for CBW use against them, although many of these are now being withdrawn. Third, while some argue that the Geneva Protocol now has the status of customary international law, others find these arguments less than compelling.

The Biological and Toxin Weapons Convention

One of the key differences between the BTWC and the other treaties banning possession of WMD is the lack of a verification regime. This is seen by many as its fundamental weakness. Efforts to strengthen the Convention at early Review Conferences included the development of voluntary confidence-building measures. A series of these was specified at the Second Review Conference, held in 1986.

At the Third Review Conference, held in September 1991, attention was focused on verification, not least because of the revelations of the scale of efforts by Iraq to pursue a BW capability. This Conference agreed to establish an 'Ad Hoc Group of Governmental Experts to Examine Verification of the BWC from a Scientific and Technical Viewpoint' — which became known as the VEREX Group. VEREX held four sessions during 1992-93 to identify and examine potential verification measures from a scientific and technical standpoint. Twenty-one such measures were identified. It was concluded that no single measure could verify compliance, but that measures in combination would have such a capability. Having concluded that, from a scientific and technical viewpoint, verification of the BTWC was possible, the VEREX report was studied by a Special Conference of States Parties in September 1994. The Special Conference agreed to establish an Ad Hoc Group (AHG) with a mandate to "consider appropriate measures, including possible verification measures, and draft proposals to strengthen the Convention to be included in a legally-binding instrument."

Two specific verification issues and one organisational one shall be highlighted here. It should be noted that this will just skim the subject — a detailed study of verification of the BTWC cannot be adequately undertaken herein.

Commercial confidentiality. While commercial confidentiality is a concern under any regime that allows inspection of commercial facilities that may be used for both prohibited and non-prohibited purposes, there are specific concerns in relation to biological industries. This is due to the use of living organisms. The concern is that if an inspector were able to remove a sample of a live organism from an industrial facility, it may be a simple task to have the organism reproduce. If that organism has been engineered to produce, for example, a pharmaceutical product, a competitor may then go into production without expensive development costs. This situation is the corollary of the issue that makes the BTWC harder to verify than the other WMD treaties, as a potential violator has only to be able to store a small quantity of live organism to retain a BW production capability.

NTM issues. The second verification issue is that of National Technical Means (NTMs). This is an antiquated term that has remained in the field through nothing more than inertia. The terminology derives from bilateral arrangements between the U.S. and USSR. These two states had a 'feel' for what NTMs were and the use of the term covered a number of sins — it should be remembered that national efforts were the only sources of information, as in the early agreements there were no formal verification provisions.

NTMs reared their ugly head during the closing days of the negotiations for the Chemical Weapons Convention in 1993. They did the same in the final stages of CTBT negotiations in July-August 1996. The question of NTMs and the test ban took a good ten days — two weeks of negotiating time — to resolve. Yet the problem need not have been there. There are many other ways to define information provided by governments to an international organisation — 'nationally-derived information' comes to mind immediately. People can easily understand what it means; limits to what sort of information could be provided may be spelled out; and, perhaps most importantly, it translates easily.[2] In both the CWC and CTBT negotiations, NTMs did not become major issues until the closing stages, catching many delegations 'on the hop'. That the same will happen under the BTWC is not only probable but practically certain. This issue needs to be tackled soon. If this is done early enough, there should not be a problem.

A new international organisation? One issue that is being discussed in the Ad Hoc Group is whether a new international organisation is required. There are good arguments, for example, about whether the OPCW — the CWC organisation — should be the BTWC organisation. Many verification technologies under the two regimes will be quite different — although both will use similar procedures. However, the costs of setting up a new organisation would be much steeper than adapting an existing one. There is a further argument

---

2. Not only is the term national technical means hard to define in English, but it is also, apparently, practically impossible to translate into other languages. This, in itself, may have been a cause of problems during negotiations.

that is rarely aired, but which is important — if there are to be two organisations, will they overlap their coverage or would a gap appear in between? International organisations are always under budgetary pressure and any visible overlap would be seen as an area where cost savings could be made and a gap may then appear. If there were one organisation — perhaps the OPCBW? — there would be no issue of overlap.

The Chemical Weapons Convention

The CWC is the youngest of the global WMD treaties. It was signed in 1993 and, following the deposit of the 65th ratification on 31 October 1996, will enter-into-force on 29 April 1997. The OPCW — the Organization for the Prohibition of Chemical Weapons — will be responsible for the Convention's verification arrangements. Inspector training continues apace and the Information Management System — yet another IMS — is being made ready for dealing with the data declared by States Parties.

Major uncertainties arise over what will happen with regard to the United States and the Russian Federation, as neither has yet ratified the Convention. In addition, the bilateral arrangements on chemical weapons and the mutual inspections of their destruction are currently the subjects of some controversy. There is a chance that both states could ratify before entry-into-force, but the timelines are tight. This leaves the OPCW in something of a planning nightmare. If the Bilateral Destruction Agreement (BDA) were to be operational and both these states were states parties to the CWC, then the OPCW's role would be limited to complementary verification to support international confidence of the CW destruction. If the BDA is not operational, and both states are states parties to the CWC, then the workload of the OPCW rises considerably. If neither state is a party, then the workload of the OPCW diminishes considerably. It is also possible — if unlikely — that one state, but not the other, could be a CWC party at entry-into-force. The ideal situation would be for both states to be parties and for the BDA to be operational. Whether this will happen is in the hands of the U.S. Senate and the Russian Duma.

Conclusions

The prospects for arms control over the next half decade or so look good, with only the CFE and the CTBT giving cause for real concern. The CFE retains its paradoxical nature: the cheapest and simplest verification arrangement is also the one that would preserve elements of bloc-to-bloc confrontation which would be seriously disadvantageous if the political climate in Europe were to cool. A new verification organisation could reduce potential bloc-on-bloc tensions, but may be financially and politically unfeasible. In other words, neither the present decentralised nor a centralised inspection system are ideal for CFE verification. It is hard to be positive at this point.

Universality as an Issue

As to the global regimes, universality seems to be a key issue. In particular, the CTBT may be next to worthless without membership of certain key states. For the other treaties controlling WMD, there are many benefits that universality brings to the enhancement of verification regimes — not the least of which is the identification of holdouts as already discussed. Also, once a treaty has achieved universal application, then the costs of verification apply to all. Industries may be regulated in the fields of arms control and disarmament — to ensure dual-use capabilities are not used for prohibited purposes. This will be particularly true in the case of the BTWC, as participants in the future verification arrangements will not want to feel that they are suffering economically because of this.

As treaties approach universality, pressures may be brought to bear on those that are not already members to join so as not to appear to enjoy an economic advantage. As much of the pressure to join may be economic, this apparent advantage of being outside the regime may become a disadvantage.

Another benefit of universality is the reduction in uncheckable allegations of behaviour outside of international norms that can be made. In a situation in which both sides are party to the treaty containing the norm, the allegation lacks credibility unless made through the organs of the treaty. As such, the allegations may then be checked. In this circumstance, there will be a disincentive to the making of unfounded allegations for political effect.

Promoting Universality: The Way Ahead?

Most non-membership in treaties is owing to inertia, insufficient political attention, or partial understanding of the benefits involved rather than because of a deliberate political decision to remain outside an international regime. It is clear that a great deal of work needs to be done to enhance universality, and that this should be done on many levels. The costs of promoting universality are small but the benefits will be large.

Chapter Five

Multilateral Control Regimes: Diverse Purposes and Congruent Processes

Gordon K. Vachon*

Introduction

This chapter explores the relationship between multilateral export control regimes and multilateral arms control regimes, not by attempting to match regimes with purposes and processes — whether diverse or congruent — but by considering certain "transitional challenges" in keeping with the overall theme of this volume. Furthermore, it approaches the issues of diversity and congruency tangentially, leaving considerable scope for the reader to consider her/his own views.

At issue here are differences in perspective: one person's perception of unwelcome "arbitrariness" is matched by another's perception of necessary "flexibility," in a complex world of capability diffusion where everything is not necessarily what it seems. These differences are increasingly being brought into play in international fora in two ways:

> *Who decides* what is to be subject to export controls, and how they will be applied?
>
> *Who concludes* whether a state party to an agreement is compliant or non-compliant with its obligations?

Perhaps a prophecy of sorts can be ventured: unless more time and imagination are devoted in the next few years to addressing these differences, multilateral export control and multilateral arms control efforts will both be diminished in their effectiveness and appeal. Nonetheless, efforts in both areas will continue in spite of their strained relationship.

---

* This chapter is an edited version of a presentation made to the Fourteenth Annual Ottawa NACD Verification Symposium, "Cyberspace and Outer Space: Transitional Challenges for Multilateral Verification in the 21st Century," 12–15 March 1997, Montebello, Quebec, Canada. The title of this presentation was chosen by the symposium organisers with a view to encouraging discussion of the relationship between export controls and arms control.

Who Decides?

The inclination of some people is to wish away the "tension" between export controls and arms control, simply by characterising it as a North-South remnant of Cold War "old think." The conventional (Western) wisdom has been, and continues to be, that export controls and arms control are complementary approaches to achieving our national and international non-proliferation security objectives. Whether focusing on nuclear, chemical, or biological proliferation, the intention is seen to be to constrain the security threats that might materialise from unhampered diffusion of certain technologies and materials. The particular concern is that states might use participation in an agreement to camouflage activities contrary to the purposes, if not the letter, of the agreement — indeed, there are instances to demonstrate that such concern is warranted.

Export controls resemble, in some respects, a prophylactic treatment to a perceived security risk, in the face of information and confidence discontinuities. Established at the national level, and likely the subject of consultation and some harmonisation among certain like-minded countries, export controls can respond more quickly — although not nearly as effectively as one would like — to proliferation concerns and intelligence in the absence of real transparency. Arms control negotiations, with their unpredictable outcomes, and agreements that are usually the product of lengthy negotiations, strive to address the same security concerns through a different set of instruments. Export control regimes and arms control agreements reflect the different timeframes and the different circumstances in which they were established and have to operate. Moreover, it has been difficult to envisage how to merge the purposes and processes of the two without raising the concern that security may be put at risk by reducing the standard of vigilance.

Others, as it turns out, do not see things quite the same way. Their opposition to Western behaviour, as represented by export control regimes, is usually still expressed in the 'developed' versus 'developing' country confrontational rhetoric of yesteryear. They highlight that export control regimes establish barriers to trade and development, as well as to their inherent 'right' to have access to technology and materials that could improve the quality of life in their countries. Yet none of these countries has taken up the challenge to demonstrate that its development has been hampered, perhaps because that would mean it would also have to demonstrate that the related security concerns were unwarranted. Many of the countries actively opposing export control regimes are the same countries which oppose effective verification of arms control agreements. More recently, there has been a shift in the thrust of the most vocal opposition, with the suggestion that existing export control regimes be replaced by reporting transfers of agreed lists of materials within the framework of multilateral arms control agreements. It remains to be seen whether this is but an astute change in rhetoric and tactics, since there has been no analysis accompanying such proposals.

In this way, the 'dialogue of the deaf' has increased in intensity in the 1990s. This involves, on the one hand, incomprehension that others do not recognise that these complementary instruments, representing different realities but working to achieve the same security goals, are desirable in their own right and benefit everyone truly concerned about security; and, moreover, are necessary in order to give effect to non-proliferation objectives in existing arms control agreements. On the other hand, there is incomprehension that multilateral arms control agreements, once concluded — and especially when comprising extensive and intrusive verification provisions, as in the case of the Chemical Weapons Convention — are not good enough. They see with suspicion the continuation of what they describe as parallel efforts purportedly addressing the same security concerns but from which certain countries would see themselves excluded, even though they are party to the related arms control agreement and subject to its verification provisions. The prospects for achieving a meeting of minds seem quite daunting if we add to this what some see as an inherent tendency toward arbitrariness in export controls and the way they are applied.

One might be tempted to say that it has been ever thus, and will ever continue to be thus, so that there is no useful purpose served by dwelling on the issue. Others see the increasing intensity of the debate as threatening to the very foundations upon which multilateralism rests: shared solutions to shared problems. As is often the case, the truth probably lies somewhere in between, since the notion of 'objective truth' applies no more here than elsewhere in the field of human endeavour where perception is conditioned by so many factors.

There have been attempts to initiate an examination of measures that could facilitate trade in dual-use materials and technology through demonstrations of self-restraint, but such proposals have not yet found favour. Similarly, there have been proposals to phase out existing, parallel export control regime approaches by embedding them in arms control regimes with (enhanced) transfer and reporting requirements to cover the same ground. But, there is so much suspicion on both sides of this debate that such proposals have not been taken seriously — whether in the context of the CWC or the ongoing negotiation of a legally-binding protocol to strengthen the Biological and Toxin Weapons Convention.

So, it seems safe to say that the 'dialogue of the deaf' will continue for some time to come. Perhaps, and *only* perhaps, a period of successful implementation of the Chemical Weapons Convention could create the conditions for bridging this communication gap, in the direction of finding shared solutions to shared problems.

Who Concludes?

Many of the same issues arise in relation to the question of who is to draw conclusions about whether a state party to an agreement is compliant or non-compliant with its obligations. There are those — perhaps drawing on their experience in the area of bilateral arms control — who argue that such 'judgements' must necessarily be a national prerogative; others argue that, if the obligations belong to the realm of an international agreement, then such judgements are a responsibility of the organs established to give effect to the agreement. In this case, as in the previous one, there are concerns as well as logic on both sides of the debate.

Among the concerns, on the one hand, is potential arbitrariness of a state in making an allegation of violation and then being unwilling to accept either the lack of evidence to support the allegation or any 'incomplete' evidence that might argue against the allegation — leading in the view of some to a complete politicisation of the verification process. On the other hand, one could be faced with a situation of a treaty organ 'voting' on a finding of compliance or non-compliance, in which group membership or alliances might be thought to come into play — also leading in the view of some to a complete politicisation of the verification process.

The United Nations Special Commission on Iraq (UNSCOM) has considerable experience in presenting factual information without stepping across the boundary into the realm of political activity. The sobering experience of Iraq's biological weapons programme, long the subject of suspicion and incomplete information — but lacking definitive proof — might not have been disclosed had it not been for the willingness of some to keep up the political pressure when others might have been willing to draw a line on the past and be content to argue that there would be greater monitoring in the future. Key throughout all of this was the unwillingness of UNSCOM to bend in its assessment of the incompleteness of the information provided by Iraq on its programme and its installations. This certainly highlights that we should not fool ourselves into thinking that serious non-compliance allegations can somehow be addressed other than in the full glare of our international political institutions. It also highlights that it is possible to envisage verification organisations functioning with professionalism and dedication, providing a factual basis for conclusions to be drawn by appropriate multilateral political authorities.

The tension between national responsibility and international organisational responsibility will undoubtedly hold challenges for multilateralism in the next few years. We must become more adept at arguing for an appropriate balance between the two. As a corollary to this debate, we must also be aware that some would wish to prevent national intelligence information from being voluntarily provided to international organisations when it relates to compliance concerns. For example, we have seen the reaction in some quarters to satellite imagery when it has been able to help substantiate concerns or provide evidence of activities denied by states. Fortunately, in this particular area the present reality of commercially-available imagery will have an impact on the factual and political contexts within which compliance questions will be examined. We must be certain that there are no obstacles to the provision of information that can have a bearing on such matters. For example, in the Ad Hoc Group negotiation of a verification protocol to the BTWC, we need pay due attention to the proposal that provisions be made for voluntary "multilateral information sharing" as a way of bringing such compliance-related information forward.

Conclusion

Multilateralism, like federalism, is a story perpetually being written. Circumstances arise that result in differences in emphasis from one period to another. We need to be prepared to challenge our established views to see if we can create new opportunities — and this also applies in such areas as export controls and arms control.

Chapter Six

Non-Weaponisation of Space: An International Imperative

F.R. (Ron) Cleminson*

Introduction

It is probably fair to say that there is no aspect of the international security equation which was not touched fundamentally by the cessation of the Cold War and the dissolution of the Warsaw Pact and of the Soviet Union. Many contend that, from a strategic perspective, the international community has been left with only one superpower. While that perception may be valid in terms of the traditional conceptualisation of international security as it pertains to quantification of "earth-bound" power projections, it is less so when applied to space. There, while the United States is pre-eminent in terms of research, investment, and infrastructure, Russia remains a world-class power given its present space capabilities — if only just barely.

Russia inherited the world's most comprehensive range of space systems from the USSR. Half a decade later, by living off the achievements of the innovative and massive progress developed during the Soviet era, Russia remains a space power of consequence. Beyond that, however, third powers such as France, Japan, China, and India are gaining rapidly in making available for international use a broad spectrum of satellite services and capabilities. With fully two-thirds of activities in space now falling within the civilian and commercial sphere, exploitation of space is increasingly dedicated toward fostering the best interests of humanity.

Outer space is the fourth environment in which humankind is active. It is unique in that while humanity has been engaged with outer space intellectually for many centuries, operational engagement has been limited to less than half a century. It is unique as well from the perspective of non-proliferation, arms control and disarmament (NACD). It is the only environment into which

---

* This chapter is an edited version of a presentation made to the Fourteenth Annual Ottawa NACD Verification Symposium, "Cyberspace and Outer Space: Transitional Challenges for Multilateral Verification in the 21st Century," 12–15 March 1997, Montebello, Quebec, Canada. The opinions expressed in this paper are those of the author and do not necessarily reflect those of the Department of Foreign Affairs and International Trade or of the Government of Canada.

weapons have not, as yet, been introduced. With the increased priority which is now accorded to the NACD processes in other environments, and particularly in view of the escalation of costs being incurred with the destruction of weapons designed for use in the first three environments, common sense suggests that the international community should be astute enough to create a mechanism by which to ensure that arms do not proliferate into this fourth environment. In other words, outer space should remain, as it is now, "weapons free."

Objective

The objective of this chapter is to provide a brief overview of certain aspects of the outer space issue as the 21st century approaches and to identify certain changes and activities relevant to the principles of the non-weaponisation of space. Paramount to that from the perspective of the international community are:

- the demise of the strategic threat;
- the international framework governing (guiding) outer space;
- definitional considerations; and,
- a multilateral approach to the issue by the international community.

Russian Activity in Space

The strategic threat and the duopoly of power which have dominated the outer space issue for more than three decades have all but disappeared. Cooperative ventures between former adversaries have come to the fore. As mentioned earlier, Russia inherited the world's most comprehensive range of space systems from the USSR. The heavy reduction in its military and civil budgets which followed have meant that Russia's overall space capabilities, particularly of its military programmes, are being consolidated and reduced further, with some corners already being cut. Production in the space industry and employment of specialist staff are estimated to have been reduced by 50 per cent or more from 1989 levels. Nevertheless, despite these cutbacks and a reduction to 25 per cent of its annual launching rate of a decade ago, Russia still accounted for one third of the total worldwide launches in 1996. Russia is a world leader in space science and sells commercial satellite imagery worldwide. Russia is also successful in the world market for launching services with priorities given to foreign launches which are hard currency generators. All in all, as Moscow attempts to reduce its reliance on other CIS states and consolidate space activities within Russia itself, it is clear that it, as exemplified by its activities relating to the international space platform, will remain a significant space power. It will promote commercial launches and collaborative projects in enhancing international visibility which, in turn, will attract foreign funding. The space 'threat' from Russia has largely disappeared.

French Activities in Space

France has achieved a high profile among others in the international community which are active in space. As one of the main players in European space activities, it ranks third — with European Space Agency (ESA) support — in terms of space-faring nations. France has maintained a diverse space programme reinforced by a scientific community of high calibre. Multilateral European cooperation is one of the major strengths and is a cornerstone of its space policy. France contributes about 30 per cent of the ESA's budget, but maintains significant cooperative relationships with the United States and the Russian Federation as well. In the coming decade, the French space industry will progressively adapt to an evolving global environment, especially in terms of space launching and earth observation.

The growing programme of civilian launches has focussed on the Ariane 4 and the Ariane 5 launchers. In terms of Earth observation, SPOT Image has been distributing high-resolution optical imagery for more than ten years. By the end of 1996, SPOT Image had captured and archived more than 4,500,000 pictures. Very recently, SPOT Image developed value-added services to meet the needs of users more precisely. It is worth noting in this context that the international community, as it assumes increasing responsibilities relating to NACD agreements, could make advantageous use of such commercial services in meeting the requirements of the ongoing monitoring and verification (OMV) process associated with significant multilateral treaties such as the Chemical Weapons Convention (CWC) and the Comprehensive Test Ban Treaty (CTBT).[1]

Other Significant Players

Japan, China, and India are significant space powers in their own right, possessing an indigenous capability in communications, remote sensing satellites, and space launch vehicles. Japan's H-II launch vehicle and its participation in the International Space Station Alpha may one day permit an independent manned space programme. China's competitive launch vehicle services and planned remote sensing satellite missions place it well on the path for achieving space power status in this area. India's IRS-IC remote sensing satellite, lifted by the independently developed PSLV launch vehicle, currently offers the best resolution from a commercial source.

1. For an interesting case study relating to the application of satellite imagery for CTBT verification purposes, see Vipen Gupta and Frank Pabian, "Investigating the Allegations of Indian Nuclear Test Preparation in the Rajasthan Desert: A CTB Verification Exercise Using Commercial Satellite Imagery," *Science and Global Security* 6:2 (1997), 101–189.

The International Community and Outer Space

Today, outer space is viewed increasingly from a multilateral perspective. The July 1993 decision by the United States to abide the "narrow" interpretation of the ABM Treaty was a unilateral gesture of singular importance in setting the agenda for progress in establishing a multilateral approach to outer space issues. The 1993 tabling by the United Nations Secretary General of a report on confidence-building measures (CBMs) in outer space provided another opportunity for the General Assembly to focus on outer space as a significant issue on its arms control agenda. The associated resolution at UNGA 48 was used to emphasise the mutuality and the complementary nature of bilateral and multilateral efforts toward preventing an arms race in outer space.

Regrettably, as the prospects for approaching the outer space issue from a multilateral perspective improve, the international community finds itself without a consensus mandate to discuss the issue. By demanding at UNGA 51 in the autumn of 1996, an outer space resolution with a negotiating mandate for the Geneva-based Conference on Disarmament 1997 session, sponsors of the resolution created a scenario in which a large number of countries found it expedient to abstain. Ironically, insistence on a negotiating mandate has ensured, in reality, that consensus on the establishment of an *ad hoc* group on the prevention of an arms race in outer space is unlikely. Had wording of the original consensus mandate adopted at UNGA 50 been retained in the UNGA 51 resolution, progress might have been possible.

The International Framework: Existing Multilateral Treaties and Agreements

The international framework said to guide and govern approaches to and activities in outer space consists of a variety of multilateral agreements. At least fourteen existing multilateral agreements which have been developed by the international community are seen to be relevant in some manner to the outer space issue. Within that package, the following five multilateral agreements are of greatest relevance to outer space.

The Charter of the United Nations (1945): The UN Charter as a basic document in the realm of international peace and security, has considerable relevance to the subject of arms control and outer space. As well, it is explicitly mentioned in several treaties which deal directly with outer space, including the 1967 Outer Space Treaty.

The Partial Test Ban Treaty (1963): Concern for radioactive fallout caused by nuclear testing was one of the strongest motivating forces behind the Treaty Banning Nuclear Weapon Tests in the Atmosphere, in Outer Space and Under Water. Of consequence here are paragraphs 1 and 2 of Article I which, in effect, make it illegal to carry out a nuclear explosion in outer space.

The Outer Space Treaty (1967): The Treaty on Principles Governing the Activities of States in the Exploration and Use of Outer Space including the

Moon and Other Celestial Bodies — commonly known as the Outer Space Treaty — is regarded as the cornerstone of international space law convention. As is evident from its full title, the Treaty establishes a basic legal framework for general space exploration and utilisation. Moreover, it marks an important step in controlling certain, though not all, arms in outer space. Its adoption brought about substantive changes in the legal regime of outer space. What before had merely been a set of non-binding guidelines now became legal obligations. This Treaty holds a central position within the legal framework governing all activities carried out in space.

The Nuclear Non-Proliferation Treaty (1968): While not specifically related to the outer space issue, the overriding significance of the NPT in terms of the overall NACD framework warrants its inclusion here. It was negotiated and drafted by the Eighteen Nation Disarmament Committee (ENDC) pursuant to the 1965 General Assembly Resolution 2028 (XX) requesting the ENDC to give urgent consideration to the problem of nuclear weapons proliferation. Article I of the NPT prohibits the transfer from a nuclear weapons state "to any recipient whatsoever nuclear weapons or other nuclear explosive devices or control over such weapons or explosive devices directly, or indirectly." It also requires nuclear weapons states "not in any way to assist, encourage, or induce any non-nuclear weapon state to manufacture or otherwise acquire nuclear explosive devices, or control over such weapons or explosive devices." This is the active prohibition.

The Moon Treaty (1979): The Agreement Governing the Activities of States on the Moon and Other Celestial Bodies is one of the most recent and little known agreements dealing directly with outer space. A Resolution which was adopted by consensus in the UN General Assembly on 5 December 1979 recommended the Treaty for signature and the Treaty came into force on 11 July 1984. The Treaty is a further elaboration of certain concepts in the Outer Space Treaty. While it does not apply to the Earth or earth orbits, the principles it contains regarding space conduct are of great interest. All of these principles, while general, are of relevance to customary space law today. As can be deduced from Annex A, few significant space-faring nations have ratified the Treaty thus far.

The International Framework: Concomitant Bilateral Agreements

There are a number of other agreements, though bilateral in nature, which have a significance in the multilateral context. Because of the guidelines developed and the experience accrued, there is a tendency to accept these agreements as part of the background which now defines customary international law. These agreements include the following.

Anti-Ballistic Missile Treaty (1972): This Treaty, initially between the U.S. and USSR, prohibits the deployment of anti-ballistic missile (ABM) defences, except for limited systems to protect each national capital and one other area (Articles I and III). The 1974 Protocol to the Treaty restricts each side to one site only. Moreover, while the Treaty permits the development and testing of fixed land-based ABM systems at selected test sites, the Parties undertake "not to develop, test or deploy ABM systems or components which are sea-based, air-based, *space-based*, or mobile land-based" (Article V (1), emphasis added). It should be noted that research is not expressly prohibited by the Treaty.

Verification of compliance with the ABM Treaty is to be provided by the use of "national technical means ... in a manner consistent with generally recognized principles of international law" (Article XII (1)). Each Party also agrees not to interfere with the national technical means of the other when used in accordance with Article XII (1). Furthermore, the use of deliberate concealment measures to impede verification by national technical means (NTM) is prohibited (Article XII (2) and (3)). This provision against non-interference with NTM has direct relevance to the law of outer space because one of the primary components of national technical means is reconnaissance satellites. In essence, this provision reinforces the legitimacy of such satellite activities. A Standing Consultative Commission (SCC) is created to deal with compliance issues and other questions relating to the implementation of the Treaty (Article XIII).

SALT I (1972) *and SALT II* (1979): These agreements limit the number of strategic delivery vehicles that the superpowers may deploy. Only one provision of these agreements directly relates to outer space. Article IX (1)(C) of SALT II prohibits the development, testing, or deployment of "systems for placing into Earth orbit nuclear weapons or any other kind of weapons of mass destruction, including fractional orbital missiles." A common understanding of this provision suggests that it does not require the dismantling of any existing launchers. This provision, however, would seem to reaffirm, and extends for these two states, the applicability of the restrictions regarding nuclear weapons incorporated into Article IV of the Outer Space Treaty.

The other features of these agreements that are of most interest from the standpoint of outer space are those relating to verification. SALT I incorporates the same provision (Article V) regarding use of national technical means as that found in the ABM Treaty (Article XII). Compliance questions are referred to the same Standing Consultative Commission (Article VI) as in the ABM Treaty. From the standpoint of OMV, however, the provisions for functionally-related observable differences (FROD) and of Externally Observable Differences (EOD) as a means of making imagery detection more effective are of note.

Definitions

Definitional considerations are important and are endemic in the consideration of outer space issues. Therefore, it is useful to establish and use, to the degree possible, appropriate definitions for the primary NACD issues, objectives, and functions relating to outer space. In many cases, confusion is created when the three following concepts are used in an interchangeable manner even though, as explained below, there are significant differences in meaning.

Military Use of Outer Space: It seems reasonable to assume that the use of outer space for arms control verification is one type of military use of that environment to which the majority of states is likely to subscribe. In SALT I, SALT II, and the ABM Treaty, the United States and the former Soviet Union accepted, within the parameters of international space law and practice, that the use of national technical means — a military use of outer space — *is* a legitimate execution of the verification process. This type of military use is inherently stabilising and therefore should be supported. Indeed, the inclusion of non-interference agreements in the Stockholm and Vienna Documents provides implicit approval at a multilateral level. Without such an application of the use of imagery-gathering satellites for verification purposes in the past, many significant international arms control agreements would not have been possible. It can be argued that other military uses of outer space, such as global communication and early warning, also have stabilising aspects.

Weaponisation of Outer Space: At the other end of the spectrum, "weaponisation of outer space" seems to refer to the placement of weapons in outer space or their use in, or from, that environment. To the best knowledge of the international community, weapons have not yet been placed in orbit on a permanent basis. It is generally assumed, however, that anti-satellite (ASAT) weapons have been inserted into partial orbits for testing purposes on more than one occasion in the past. Apparently, the trajectories of intercontinental ballistic missile (ICBM) systems have not been interpreted to involve the weaponisation of outer space. The important distinction between weapons placed in outer space, weapons which only transit outer space on the way to their targets, and weapons based elsewhere which are used to attack targets in that environment, is often blurred in discussions. Nevertheless, for our purposes, the recent announcements by the United States and the Russian Federation regarding their adherence to the conventional understandings of the ABM Treaty are encouraging and have contributed significantly to the building of confidence in this area.

Militarisation of Outer Space: Finally, between the "military use of outer space" — which seems likely to be acceptable to most states — and the "weaponisation of outer space" — which is generally opposed by most states — falls the concept of "militarisation of outer space." While this term is particularly vague, it appears to imply less of a military presence than "weaponisation" but more so than "military use." The proceedings and working papers of the CD are replete with references to the "prevention" of outer space militarisation or to the "problem" of "non-militarisation" of outer space. Other states have made reference to the need for the "demilitarisation" of outer space. To some states, "militarisation" seems to be used in almost the same sense as "weaponisation" — that is, to refer exclusively to weapons. Other states seem to use the term so as to include any military use of outer space.

If outer space were to be militarised, it would seem logical to assume that some sort of military authorisation would be necessary before payloads could be placed in orbit. That is not the case today. Therefore, we might assume that, while military use of outer space is an accepted practice, outer space itself is not "militarised." Probably most would agree that it should not be.

Terminology

While the problem of definitions must be addressed, terminology relating to outer space issues, and venues within which to discuss and negotiate issues, is also potentially troublesome. For example, many expressions used today in the CD have their origin in the deliberations leading up to the Outer Space Treaty of 1967. The venue within which much of this debate took place, and continues to occur, is the Committee on Peaceful Use of Outer Space (COPUOS). The mandates of COPUOS and the CD are distinctive and should be seen as mutually reinforcing. Nevertheless, their responsibilities should be clearly delineated to avoid overlap, particularly since the environment to which each mandate applies is the same. Their pertinent terminology and definitions are, therefore, closely interrelated. This is an area in which continued and increased cooperation is not only desirable but is, indeed, necessary. The significant expansion of commercialisation in space, without an adequate regulatory environment — i.e., no central authority to adjudicate — could lead to misunderstanding and conflict.

Multilateral Approach

The international community can gain advantage in its associated activities by approaching outer space from two perspectives. The first is as an environment which should be kept free of weapons. The second is as a base from which remote sensing systems might operate in support of the monitoring and verification of compliance with agreements relating to earthbound commitments. In both cases, remote sensing by satellite in a space-to-space or space-to-ground mode is essential.

Space-to-space: In terms of verifying that outer space remains weapon free, one might expect that the determination of the exact function of an unknown spacecraft would be exceedingly difficult. In fact, the ability to verify an arms control treaty relating to outer space is, at best, theoretical. No known operational space-based weapon system has yet been deployed in space. Thus, there is still a considerable amount of uncertainty as to how these systems would be configured for optimal performance. However, the high degree of optimisation inherent in the design of all spacecraft and in their orbital parameters, together with the nature of signals to and from the spacecraft, provide significant data as to their function.

Clearly, to the extent that form follows function, visual images of the spacecraft are highly determinate of distinctiveness in configuration. For example, if images can be acquired in the thermal-infrared region, important data can be derived regarding the energy balance and energy utilisation of the unknown spacecraft. The high degree of optimisation inherent in the design of all spacecraft and in their orbital parameters, together with the nature of signals to and from the spacecraft, can provide highly significant data as to that spacecraft's function. The fact that military and related space missions have orbits clustered in limited portions of space simplifies the problems of space-based observation. The observations required to fulfil such a mission could be carried out with a group of sensors, including a visible-light imaging system, a thermal imaging system, a receiver to measure radio frequency signals, and radiation and chemical sensors. Such observations would likely be undertaken by remote sensing satellites, ground-based systems and, perhaps, by launch site inspections.

Space-to-ground: In terms of monitoring and verifying compliance with existing arms control agreements, there are now at least a half dozen significant multilateral NACD agreements and treaties either in existence, in the implementation stage, or under active negotiation or discussion. In 1980, a dedicated satellite system might have been thought necessary. Today that is no longer essential. Suffice it to say that commercial and other open source overhead sensing can now provide an alternative and cost-effective mechanism in support of multilateral NACD — OMV.

Within the next twelve to eighteen months, a commercial market offering imagery with ground resolved distance (spatial resolutions) as small as one to two meters in the visible spectrum will come into existence. Multispectral imagery from space at somewhat poorer resolutions will follow shortly thereafter. Synthetic aperture imagery is now becoming available through the use of Canada's RADARSAT. The emerging commercial market for date and image products from satellite-based sensors should be assessed in terms of making positive contributions for NACD purposes.

Outer Space Today

In the United States, the pre-eminent space power, there continues to be a considerable debate over the timing and nature of a national missile defense (NMD) system and the extent to which outer space is required in terms of a basing mode. While both the Clinton Administration and the U.S. Congress agree on the requirement for an NMD, there are major differences regarding the scope and timing of the threat. Some consider that Alaska and Hawaii will be within the estimated range of the North Korean Taipo Dong 2 ballistic missile, currently under development. Others believe that there will be no threat for at least a decade, and that the intelligence community has the capacity to identify the threat in sufficient time to be able to deploy an NMD system. Still others believe that the possibility of an accidental or unauthorised launch from Russia or China justifies the deployment of an NMD system now.

From a multilateral perspective, the threat of a superpower arms race in outer space is seen to have effectively disappeared. However, the concept of placing weapons in space, ostensibly for defence purposes, has not. An arms race in outer space in the future may be more the result of competition between corporations, think-tanks, and pressure groups rather then between states.

Beyond the standard security and stability considerations, there are a number of non-traditional security challenges which might become serious threats if not identified and rectified at an early stage. The commercialisation of space activities, already addressed, could pose dangers for peaceful uses of outer space because of the absence of a central regulatory authority. With the exception of the ITU, which deals with frequency regulations, there are few examples of institutions accepted within the framework of customary usage which might be seen to function as a central focus. One such area relates to the non-interference with NTMs which is included in bilateral strategic arms treaties, as well as in multilateral agreements such as the CFE Treaty, the Vienna Document, and the CTBT. Internationalisation of such agreements might be considered in codifying these and other principles defined in the ABM Treaty and elsewhere. A survey of national positions might be justified to identify areas where there are common views.

In this connection, for example, Canada, Russia, and China are included in a large number of states which appear to share a similar concept in terms of outer space and arms control. These countries have declared support for the policy of maintaining outer space as "weapon free". All three have supported with different degrees of enthusiasm the concept of using space-based remote sensing technologies as part of an international monitoring and verification system in support of compliance with multilateral treaty obligations. Clearly, there are areas in which a majority of states could find consensus in the formulation of a broad international framework of understanding. Perhaps it is time to call for a review conference on the 1967 Outer Space Treaty from which a consensus could emerge for a multilateral approach to ensuring that outer space remains a weapons-free environment into the 21st Century.

Conclusions

- The concept of maintaining outer space as a weapons-free environment, which is supported by the majority of states, including Canada, should be recognised as an area of prime multilateral importance. Codification of the principles ensuring that space remain weapons free could be developed in a review conference aimed at updating the 1967 Outer Space Treaty.

- A survey of national positions on basic principles building upon the existing treaty framework would provide a useful "multilateral" basis from which to ensure peace and security in outer space.
- Space-based remote sensing can contribute effectively as part of a monitoring and verification package for significant multilateral agreements.
- There is a significant role for the United Nations to play as facilitator in establishing cost- effective, multilateral services in the area of non-proliferation, arms control, and disarmament.

Chapter Seven

The Proliferation Challenges of Cyberspace

David Mussington*

The Emergence and Structure of Cyberspace

What is Cyberspace?

The emergence of global information networks as a venue for economic and political activity carries with it an unfortunate consequence: the potential use of these same networks for hostile political and/or criminal purposes. Because the emerging global information infrastructure (GII) is an amalgam of different national, international, public, and private communications and computing systems, there is no single authority or agency with a mandate for protecting common interests in enhanced network connectivity.[1] Instead, cyberspace — the increasingly interconnected network of computers and telecommunications systems serving businesses, consumers, and governments — is governed by an *ad hoc* framework of increasingly obsolescent legal rules, public and private telecommunications services providers, and a rapidly growing user base.[2]

Physical and Virtual Layers in Cyberspace

Cyberspace comprises two linked but conceptually distinct layers: (1) the physical infrastructure of copper wire, microwave, satellite, and fibre optics networks, switches, routers and bridges, and servers that make up the backbone of the GII; and, (2) a virtual or information layer, comprising the software protocols making up network operating systems and information in the form of data flowing in networks that is time-sensitive, economically valuable, and/or politically sensitive.

---

* This chapter is an edited version of a presentation made to the Fourteenth Annual Ottawa NACD Verification Symposium, "Cyberspace and Outer Space: Transitional Challenges for Multilateral Verification in the 21st Century," 12–15 March 1997, Montebello, Quebec, Canada.

1. For an early formulation of these relationships, see my "Throwing the Switch in Cyberspace," *Jane's Intelligence Review* (July 1996), 331–334.
2. There is a danger that scientific users of information networks will be displaced by more lucrative commercial applications in broad system design decisions. This would be a tragic event, given the increasing importance of scientific collaboration on the Internet. See Herb Brody, "Wired Science," *Technology Review* (October 1996), 42–51.

These layers require different treatment if threats to them are to be comprehensively understood. It is at least possible that measures taken to protect the physical infrastructure of information networks may have a negative impact on the flow of information within the networks themselves.[3] In turn, actions taken to secure the economically and politically sensitive flow of information on networks could undermine their physical security. In both cases, however, the issues determining whether the physical and virtual layers of cyberspace develop concurrently, derive from the characteristics of the two environments. This chapter offers a framework for understanding emergent network vulnerabilities, and for understanding non-network vulnerabilities that are exacerbated by the enhanced availability of information on global networks. This analysis highlights the difficult balance that must be struck between protecting critical networks, and maintaining the open access and global scope that give these networks their special value.

The term 'cyberspace' is used in this chapter because it encompasses a broad range of terms used to describe an emerging family of wide area networks linking advanced computing and communications infrastructures together.[4] Because of its origins in science fiction, cyberspace is held by some observers to be little more than a fictional and 'hype-ridden' phenomenon, promoted by the popular media, and likely to collapse once the public's fascination with it wears off. However, this is unlikely to be the case.

The emergence of national and, increasingly, global information networks is being underwritten by influential governmental and economic interests seeking to define new markets, and to protect hard-won technical competencies.[5] As the extension of the 'personal computer revolution', the growth of the Internet is likely to cause a significant shift in patterns of economic, and eventually, political activity.[6] This environment will likely contain significant

---

3. Computer Science and Telecommunications Board, National Research Council, "Application Needs for Computing and Communications," *Computing and Communications in the Extreme: Research for Crisis Management and Other Applications* (Washington, D.C.: National Academy Press, 1996), 10.

4. The term was first found in a science fiction novel entitled *Neuromancer*, by William Gibson. In that context, the term described ubiquitous network and direct brain-computer interfaces. While such applications are not on the immediate horizon, the societal and vulnerability consequences of network dependence are beginning to make themselves felt.

5. A significant analog to the growth in commercial network applications is the increasing importance of information systems to national defence. These systems are not the primary focus of this analysis for a single reason — they are following, not leading — the deployment of broadband commercial networking systems. For a good summary of the technology security issues raised by these developments, see Ian Anthony, "Transfers of Digital Communication System Technology," *SIPRI Yearbook 1996: Armaments, Disarmament and International Security* (London: Oxford University Press, 1996), 552–559.

6. Roger C. Molander, Andrew S. Riddile and Peter A. Wilson, *Strategic Information*

political and social conflicts and, of course, forces favouring cooperation and conflict resolution.

The study from which this chapter derives focused on the problematic aspects of global networks — those that may empower new actors to threaten important societal infrastructures. The emerging threat is of two types:

- problems arising from the increased availability of information useful to those interested in procuring materials and technologies for the creation of weapons of mass destruction; and,
- problems created by the spread of information warfare capabilities to state and non-state actors that oppose Western interests.

Information weapons may allow terrorists — or others acting on behalf of states — to create new capabilities for blackmail or intimidation. In turn, the prevalence of complex technological systems in society — such as electric power plants, communications networks, and air transportation systems — extends this vulnerability beyond the military sphere, to include broad segments of civil society. Responses to these potential vulnerabilities must take into account the unique characteristics of cyberspace. The first section of this chapter focuses on the physical and virtual domains within cyberspace. This chapter relates the features of the network environment to the possible empowerment of new actors who may pose threats to critical societal technological systems and information resources.

Two Domains in Cyberspace

Because cyberspace has two conceptually distinct domains, threats to information infrastructures can be discussed in terms of their impact on both information flow, and decisions to deploy the new network technologies making up the GII.[7] The physical and virtual domains within cyberspace each possess characteristics that suggest that they are potentially attractive targets for attack by those wishing to undermine the security — or credibility — of the GII as an environment for political and economic activity.

Virtual Layers Issues

The information, or virtual, layer of cyberspace comprises both the software allowing the physical infrastructure of the GII to function, and the data and applications which create and transmit economic and political information among users. The three characteristics or concepts that describe the virtual layer of cyberspace are:

Warfare: A New Face of War (Santa Monica, RAND/National Defence Research Institute, 1996).

7. Mussington, "Throwing the Switch in Cyberspace," 332.

- the requirement for authentication — in terms of both users and information — of the data transmitted over a network;
- the necessity that information be transmitted through networks in a rapid manner for time-sensitive economic purposes and critical public safety applications; and,
- the desire that access to the network be widely distributed, while at the same time preserving the authenticity of information and identity in cyberspace.

Authentication. This requires that authorised users and creators of information on critical networks have confidence in their ability to determine — through the use of network functionality — their mutual identities. On the Internet, authentication schemes dependent upon strong encryption seek to ensure against unauthorised use of network-carried information or applications, examples of which are: initiating transfers of funds using inter-bank funds transfer systems; the closure/opening of valves within oil and natural gas pipeline networks; and, the erroneous detection of aircraft — or the incorrect identification of some — by an air traffic control system.[8]

Timeliness. In applications that require the use of real-time information delivery, any system breakdown or fault can impede the flow of information and compromise system performance. Examples of real-time dependent applications include: financial data and electronic funds transfer systems; "just in time" inventory control systems; and, medical information systems. A subtle degradation in system performance may have significant consequences for network users. Information "outages" for such systems are intolerable, as are real-time interruptions in network control and management. Examples of network control applications include: SCADA networks — remote control and maintenance systems for oil and natural gas pipelines; digital switches underlying the public switched telephone network; and, regional and national air traffic control systems.[9]

Access. This issue differs from authentication in that it refers to "ease of use" issues deriving from the actual systems and procedures deployed in real-world networks. For many applications, restricting access undermines economic viability — i.e., as may be the case in Internet commerce. This is preeminently an issue of interface design, where the imposition of user authentication schemes begins to impede the free flow of information, or the use of network applications. Ease of access clearly has a different meaning in different segments of the GII, be they secure funds transfer systems, ordinary e-mail, or remote control and maintenance applications. Where different network applications use the same physical infrastructure — i.e., the public switched tele-

---

8. Deborah Russell and G.T. Gangemi, Sr., *Computer Security Basics* (Sebastopol, CA: O'Reilly & Associates, 1992), 124.
9. *Computing and Communications in the Extreme*, 31.

phone network — different users of information resources may become vulnerable to impairment from a single set of vulnerabilities.[10]

Physical Infrastructure Layer Issues

The physical part of cyberspace is made up of the copper wire and fibre of the public switched telephone network, microwave and satellite communication systems, digital switches and routers allowing for isolated networks to be interconnected, and the individual servers and computers which make possible particular information-dependent applications. Three concepts describe the security issues deriving from the physical network infrastructure:

- redundancy — the existence of multiple and duplicative network links between network nodes;
- fault tolerance — the ability of a system to manage impairment to system function short of a catastrophic (complete) failure of critical functionality; and,
- physical security — the robustness of critical links and servers against destruction through accident or deliberate assault.

Redundancy. This is, perhaps, the central attribute of the Internet, stemming from its origins in the original ARPANET system designed to ensure communications connectivity during and after a nuclear war.[11] Beyond the packet-switched architecture of the Internet, communications networks use different modalities for the transmission of information. Among the systems in use are: microwave networks; satellite and terrestrial fibre/copper wire telephone systems; conventional radio broadcast networks; and, cellular networks.

These varied systems increase the robustness of the GII to technical failure, but it remains possible that shared infrastructures supporting different parts of the GII may subject the system as a whole to the vulnerabilities of the weakest segment. It is possible, for example, that weakness in the public switched network (PSN) could propagate across the GII in novel and unexpected ways if software or other related operational factors begin to impede the operation of switches and routers. Such a situation arose on 15 January 1990 when AT&T's long distance network suffered a catastrophic failure in the United States. Flawed system software in some of AT&T's digital switches, along with operational rules that mandated that all switches receive software upgrades at the same time, created a situation many experts said could not occur. User confidence in such networks requires that multiple redundant pathways be built into both individual components of the GII and into the

10. *Computing and Communications in the Extreme*, 47.

11. In many ways system security concerns for a military application are different to those of civilian systems. The working out of these differences takes place in the economic and political decision-making underlying GII development. See Russell and Gangemi, 28.

network as a whole. In the case of cyberspace, the varied sources of hardware and software connected to shared communications infrastructures open the possibility of nonstandard equipment causing network problems.

Fault Tolerance. This concept describes the reconstitution and robustness of a system in the face of software and hardware errors, unanticipated system state changes, and unauthorised interference with a network through physical assault. The question of how the network manages divergences from expected system behaviour is critical. If a system degrades gracefully, some system functionality is maintained even if the overall system architecture no longer operates within expected parameters. If the system does not manage small flaws well, then its vulnerability to the failure of a few critical components is heightened. The PSN and the Internet are said to degrade gracefully because of the number of alternate pathways through which messages can pass if critical switches and routers are no longer functioning.

Because the virtual and physical layers of cyberspace interact closely with one another, synergies — or unpredictable combinations of effects — exist in the combination of software and hardware that have the potential to:

- render cyberspace much more resistant to sabotage and unauthorised penetration than is commonly assumed; or,
- create unanticipated vulnerabilities, where the failure of software or a hardware component or subsystem may propagate across networks in unanticipated ways.[12]

Factors that influence this issue include: the frequency and nature of software upgrades to critical switches and subsystems; the nature of the technical standards underlying key network technologies — whether they are strong or weak; and, the extent to which economics influences network design. Concerns for maintaining system functionality encourages the building of redundant, fault-tolerant systems for critical applications. The public telephone network is the best example of this type of system. As the inheritor of the current communications infrastructure, the emerging cyberspace environment cannot help but be influenced by extant design disciplines and infrastructure resources.

Physical Security. To what degree are critical servers and communications links secured from physical damage? What is the defining method for protecting these essential elements of the physical infrastructure of cyberspace? And, is an accident viewed as the most likely threat to system integrity, or is deliberate attack viewed as a serious possibility?[13] The protection of physical

---

12. See *The Hacker Crackdown* by Bruce Sterling for a discussion of the ways that a system fault can have unpredictable and multiplicative consequences. Bruce Sterling, *The Hacker Crackdown: Law and Disorder on the Electronic Frontier* (New York: Bantam Books, 1992), "Part 1: Landslides in Cyberspace."

infrastructure requires investments in barriers to entry, redundant and fault-tolerant systems that protect networks from limited assault, and the diffusion of encryption and authentication systems that can foil "casual" penetration attempts by hackers and "experimenters."

Each of these factors impacts the choices available to system designers and analysts concerned with protecting critical information resources. Networks, by their very nature, are both distributed computing systems and communications-centric entities. This means that maximising the usability of the network is a key system value, and not one that can be easily compromised in favour of greater security. Increasingly, internationally distributed networks open different systems to the weaknesses of other subsystems within the wider network environment. Regulatory responses to network security lapses must thus have an international component if key information networks and applications are to be adequately protected.

Types of Network Vulnerability

Three major types of network vulnerability are likely to affect the emerging GII:

- network failure due to technical breakdown;
- the theft of information from the network; and,
- the targeting of information resources for falsification or destruction.

Each of these threats to network functionality requires governmental responses. The key issue in each case is the level of severity of negative effects from the problem's occurrence, and the appropriate policy response from government agencies.

Network Failure Due to Technical Breakdown. The growth in connectivity between previously separate networks means that minor vulnerabilities have the potential to propagate across a wide geographic area — affecting more users than might be readily anticipated beforehand. An example of such an event was the failure of the AT&T long distance telephone system mentioned above. In that case, 'buggy' software upgrades installed at digital switches on the long distance network succeeded in cutting data links and communications between three large airports on the U.S. East Coast — New York's Kennedy and La Guardia Airports, and New Jersey's Newark International Airport.[14]

---

13. The operating security management "metaphor" for information dependent infrastructures varies widely across systems. Air Traffic Control systems focus on accident or system breakdowns as the predominant source of breakdown risk. Oil and Natural Gas pipelines are similar in this way, but have more latitude in allowing fault tolerant systems to handle problems than do air traffic controllers who are necessarily much less tolerant of system outages or operations falling outside of accepted parameters. See National Research Council Report, *Understanding the Risk: Informing Decisions in a Democratic Society* (Washington, D.C.: National Academy Press, 1996).

14. Sterling, *The Hacker Crackdown*, 37–38.

These events caused a reciprocal slowdown in connecting air traffic in both the United States and Western Europe.

The added vulnerability that flows from greater connectivity also makes it difficult to differentiate accidents from deliberate attacks on critical systems. More than anything, the AT&T incident exemplifies the relationship between access and authentication mentioned earlier. In an effort to increase the functionality of its long distance network, AT&T introduced new vulnerabilities into its systems because of flawed software in digital switches. Thus, measures taken to enhance the accessibility of the network to users — i.e., clearer voice lines, faster connections, and lowered error rates in data transfers — actually reduced the robustness of aggregate system security.

Theft of Information. The most commonly discussed vulnerability of information networks is that of the theft of information with economic, political, or national security value. Since the early 1990s, a number of arrests have been made in the U.S., Germany, and elsewhere of 'hackers' subsequently convicted of illegally penetrating secret government computing facilities. Attacks against economic targets are also of increasing concern, especially following the revelation in the early 1990s that the French Intelligence Service was targeting foreign — principally Japanese and U.S. — businessmen for industrial espionage.[15] In this case, economic and political goals combined with one another, revealing the difficulty of separating complementary motivations.

At the end of the Cold War, many intelligence agencies have apparently shifted to economic intelligence gathering from their formerly military missions. This means that national and international policy must decide between a focus on counterespionage, and one more directly targeting criminal activity in the deployment of resources to counter system vulnerabilities. In 1990, a Russian man affiliated with Moscow organised crime groups managed to penetrate the electronic funds transfer systems of a major U.S. bank, stealing $10 million. While most of the funds were recovered, the successful penetration of secure and private information networks highlights the increasing interconnectedness of the world's networks. The more that these networks share physical infrastructures — and, increasingly, come to resemble each other in their virtual applications — the more vulnerable they may become to attacks that exploit small weaknesses in security preparations.

Targeting of Information Resources. Another source of cyberspace vulnerability is the potential compromise of secure and sensitive information held within networks. Theft of this information may have significant value, as the discussion above outlines. A more subtle means of attack might be to systematically compromise the timeliness or accuracy of critical data used in some

15. Winn Schwartau, *Information Warfare. Chaos on the Electronic Superhighway* (New York: Thunder's Mouth Press, 1995), 272–273.

economic, regulatory, or emergency medical application. Delays in the delivery of information on price movements in international stock and commodities markets could seriously disadvantage corporations and countries operating with impaired computing facilities. Similarly, the impairment of data on system behaviour in the SCADA system controlling natural gas pipelines could create potentially serious network management problems. In a worst case scenario, physical damage would be done to infrastructures if real-time data management systems were attacked. Lastly, a logistics system dependent upon computerised records of the disposition of physical goods and personnel might be paralysed by the introduction of erroneous information into data systems supporting military deployments. In the future, "just in time" inventory control methods may open up new vulnerabilities in governmental information management.

Understanding Vulnerability and Its Consequences

The discussion above outlines some of the vulnerabilities that characterise emerging information networks. What is the appropriate policy response to these weaknesses? The problem of detecting system flaws lies behind many of the problems outlined above. Put simply, if a system has not failed in the past, it is assumed to be "time-tested" and unlikely to fail in the future. Unfortunately, networked systems are vulnerable to the characteristics of the last piece of equipment or subsystem added to the system. This means that the exact impact of flaws in system software or in connectivity-related areas is, in many ways, inherently unpredictable. Networks are probably subject to "normal accidents," a term coined by Charles Perrow to describe system interaction effects that are unpredictable due to the complex nature of software and hardware relationships in large technical systems.[16]

The complexity of networks holds a troubling implication for arms control and export control issues. The ambiguity of system behaviour intimated by Perrow's insight means that new actors are progressively empowered with the potential to do widespread disruptive damage to critical networks. It is at least theoretically possible to reach almost any computer connected to the public switched network from any other. This multiplication of access points to the virtual layer of cyberspace means that new actors, non-state or state, criminal or terrorist, may gain the ability to stage complex and subtle attacks on critical and economically valuable activities taking place throughout cyberspace. It also means that these entities may be able to use the information available on the Internet to threaten far more destructive behaviour in the proliferation of weapons capabilities in the WMD category. The next section of this chapter expands on the nature of the actors that may be able to use cyberspace to

16. Charles Perrow, *Normal Accidents: Living with High Risk Technologies* (New York: Basic Books, 1984), 17.

attack civil society, and outlines the new — and not so new — nature of their motivations for doing so.

New Actors and Empowerment in Cyberspace

The broader availability of information and tools for launching "combined arms" — both disruptive and destructive — attacks on critical network systems are creating the potential for new actors to emerge as significant players in political conflicts where previously only states were of great significance. These actors can be characterised in terms of both their phenomenology and their goals. These goals can be discussed in terms of the target of an assault, the nature and scope of destruction or disruption, and in terms of the unanticipated weapons effects that may result.

Figure 1: Characterising Threats and Actors in Proliferation Worlds

	Actors					
Goals	Religious Extremists	Intelligence Agencies	Terrorists	Transnational Criminal Organisations	Rogue States	Individuals
Extortion			X	X		X
Promoting Anarchy	X		X			X
Other Ideological Goals	X	X	X		X	X
Geostrategy		X	X		X	
Economic Gain		X	X	X	X	X
Furtherance of Religious Beliefs	X		X		X	X

While this matrix of actors and goals is necessarily simplified, it suggests that the number and variety of actors who might take advantage of network vulnerabilities for political purposes are quite large. As is true in many other areas, after the Cold War, economic, sectarian, and religious motivations for group action are likely to be significant organising principles for new actors in this hypothetical networked world. Coalitions among the various actors may also be possible, making the determination of authorship of particular attacks — and of the underlying motivations behind them — increasingly difficult to achieve. The potential thus exists for "piling on" — in a way resembling the "taking of responsibility" statements released by terrorists in the aftermath of significant destructive events. Identifying the responsible party in instances of terrorism is central to governmental policy-making. Ambiguity is correspondingly likely to impede effective and rapid reactions by governments — at least those tailored to the actual perpetrators of information warfare attacks. The anonymity of network attacks — or, rather, the difficulty of tracing the geographical whereabouts of perpetrators — makes 'hacker' activities aimed against critical information infrastructures a likely means of terrorist exploitation.

Already, visible trends in technology diffusion and computer research point to a potentially significant expansion in the number of actors in the proliferation area.[17] A number of key shifts in technology and politics will help to reinforce this trend. The long-anticipated rise of non-state actors as bearers of serious proliferation threats will likely emerge soon after the year 2000. Precursors of these new actors are already visible, with the Aum Shinrikyo cult's use of chemical agent in the Tokyo subway being the most serious and significant example.[18] This group is doubly significant because its activities highlight the increased difficulties facing verification/detection agencies charged with policing the possession of potentially dangerous compounds or weapons. Two facts are key to Aum Shinrikyo's activities.

- The cult actually launched three attacks, only one of which was ascribed to them in the first instance. The other incidents were blamed on accidents, or safety violations associated with legitimate uses of chemical compounds.[19] The cult used the Internet to obtain information on bomb making and to coordinate its activities across geographically significant distances — with adherents as far separated as Japan, Russia, and South Korea.
- Each of these factors — the Internet as a coalition support and the use of pre-tests by non-state actors to "hone" their skills — are likely to recur if such groups are active in the future.

The continuing spread of dual-use technologies means that the "ambient technological level" of proliferators will be that much higher, further empowering new actors in proliferation. Each of these factors supports the insight that these new actors will pose proliferation threats to national and global security in coming decades. Reclassifying the goals of these new actors into smaller categories yields the following array of proliferator objectives:

---

17. Anthony, "Transfers of Digital Communications System Technology," 554.
18. "The Tokyo Nerve Gas Attack and CBW Terrorism," in *SIPRI Yearbook* (1996), 701–704.
19. *Ibid.*, 702.

Figure 2

Goals	Actors					
	Religious Extremists	Intelligence Agencies	Terrorists	TCO's	Rogue States	Individuals
Economic Gain or Extortion		X	X	X	X	X
Ideological or Religious Goals	X		X		X	X
Geostrategic Objectives		X	X		X	

Proliferation threats in this environment change shape to resemble better understood — though not well countered — terrorist or criminal activities. As pure (ideal) types, the different cells suggest that the future will be at least as dangerous as today with serious proliferation dangers deriving from the complex combination of disruptive and destructive instruments. This combination may produce new instruments of utility to terrorists and foreign states wishing to avoid a clear 'paper trail' proving their responsibility for network disruption and damage.

The ambiguity of authorship in network penetrations or disruptions means that these events may be designed to resemble anticipated accidents or instances of human error. The detection/verification task in this domain thus becomes exceedingly difficult. Differentiating normal network behaviour from the deliberately induced degradation of key system functions requires that methods of investigation and verification be designed with a close eye on the nature of the networks that are under threat. Governmental concerns with political-military security merge with corporate and societal concerns with privacy and the security of intellectual property to exacerbate the difficulties of making policy in this area. Criminals or terrorists may take advantage of this yet unstructured environment to attack key information infrastructures and the accuracy and credibility of information held by government and private sector entities.

New Actors and Government: Capabilities versus Regulatory Effectiveness

In traditional areas, states enjoy an unquestioned superiority of legal and material resources relative to any other type of actor in domestic civil or international society. Asymmetries in intellectual or technological capabilities are commonly examined within an institutional legal order that advantages governments and legitimated authorities as the ultimate arbiters of permissible behaviour. Terrorism and guerilla warfare are both classic challenges to governmental preeminence, and the new actors in the cyberspace environment borrow some of their effectiveness from 'unconventional' warfare areas. The exact nature of new capabilities available to actors with network access will probably vary considerably. Network access alone does not equate to the ability to manipulate information systems and information content in ways which impact the public sphere. Because the relationship between network expansion, individual empowerment, and governmental regulation of computing and telecommunications is complex, many observers have come to inflated and unrealistic conclusions regarding the 'end of the nation-state,' and the 'transcendence' of Cold War patterns of interstate relations. One need not go to these lengths to posit that non-state actors have perhaps the most to gain — in terms of disruptive capabilities — from the spread of new network connectivity.

The opportunity for new coalitions of non-governmental organisations (NGOs) to share information and mutual support offers these groups an impressive new resource for influencing government policies.[20] In a variety of areas, from environmental protection to conflict resolution of sectarian and religious disputes, NGOs have become increasingly active as important players in dispute settlement. In a democratic context such activities are accepted — if not welcomed — as legitimate political participation.

However, influencing the discourse on a political topic is not the same as foreclosing through intimidation or threatened action, the options of a national government. The same network characteristics which assist protest and advocacy groups in making their voices heard also give to less peacefully inclined actors the capability to attack common infrastructures — most directly, of course, the Internet itself — as a means of fostering change in public opinion or government policy.

It is in this sense that capabilities for disruption of civil society are increasing in a manner directly related to the expansion and interconnectedness of global information networks. As critical information and communications networks come to share a common physical infrastructure, they also come to share vulnerabilities to groups wishing to disrupt them. In this sense, again, disruptive capabilities are becoming more widely distributed at the very time that regulatory frameworks for network governance are becoming fragmented under the pressure from demands for deregulation from the private sector.[21]

---

20. John Arquilla and David Ronfeldt, "Cyberwar is Coming!" *Comparative Strategy* 12:2 (April-June 1993), 141–165.

21. The importance of business and societal concerns over network governance is focused by the fact that, for the most part, the private sector owns the information infrastructures. This means that the government must assert a sovereign interest in an area where infrastructures are becoming increasingly international, a reality that makes policy-making very difficult. See *Computing and Communications in the Extreme,* 40.

It follows, then, that the resource capabilities of governments in regulating networks in civil society are not easily translated into effective regulatory choices for guiding network development. Governments face an uphill battle in reasserting their primacy in the development of these new infrastructures. Rather than being tied to the computational and remote login capabilities of even the most powerful personal computer, network users enjoy theoretical access to the most powerful computing and communications instruments in the networked world. Factors such as access restrictions, authentication of users, and the timeliness of delivered data, all impinge on the way that network access is administered. Each of these factors has been a target of 'hacker' attack in the past. It is to be assumed, then, that emerging networks will also be subject to attempts at illicit penetration and compromise.

The vulnerabilities of these networks thus constitute a new kind of proliferation problem. The proliferation of means of attack is matched by a parallel expansion in the number of theoretical access points from which an attack might be launched. Differentiating attacks from accident, or human error, may prove difficult. A recent example will make this point clear. In August 1996, fifteen states in the U.S. Southwest suffered serious electrical outages. Five weeks prior to this, six of these states had been 'blacked out' by what was later traced to a short in a transformer station caused by a tree limb touching an electrical wire. The second incident provided critical information that helped managers at the Bonneville Electric Power Company to characterise the first incident. Programmed system behaviour following an accident — in this case automatic shut downs of power substations and the imposition of a brownout to a large segment of the user base — produced an outcome that many analysts had previously discounted as extremely unlikely. Under tense political conditions, perhaps in a time of civil unrest or international crisis, such occurrences could produce panic among the public, and create political pressure on leaders to address an apparent weakness in national security.

The shape of the proliferation world formed by the increased disruptive capabilities of potential proliferators and the decreased effectiveness of governmental oversight is rather stark. The accelerating pace of change in network technologies means that governmental responses are particularly difficult to design, and the still largely hypothetical nature of the dangers means that public pressure to respond creatively to this problem is unlikely in the absence of a significant information infrastructure crisis.

Because most information infrastructure attacks are likely to be of a disruptive rather than destructive character, it is possible that proliferator behaviour will be channelled toward a domain where the developed taboos against the use of weapons with widespread effects are the least developed. This potential, along with the progressive empowerment of non-state groups relative to government authorities, may create the potential for novel combinations of disruptive and destructive instruments of use to those entities intent upon altering the policies and/or behaviour of national governments through deterrence or intimidation. The next section offers a framework for understanding the relationship between verification methods and information warfare, and offers some guidelines for approaching this rapidly evolving verification challenge.

Verification Methodologies and Information Warfare

If information warfare tools are taken seriously as weapons of potentially 'mass disruption', then verification schemes designed to ensure against such activities are critical to international peace and security. The disruption of critical information networks could form part of a "combined arms" attack against civil society, where destructive means could be used in tandem with disruptive infrastructure assaults to intimidate the public. Verification in this context involves the maintenance of surveillance over networks on a day-to-day basis, so as to ensure that a baseline situation of "non-attack" is well understood. Clearly, international coordination of such efforts is mandatory, given the increasingly global nature of network infrastructures. The information-rich nature of cyberspace creates the parallel danger, however, that information useful to fabricating weapons of mass destruction may become more readily available on the Internet.

Controlling access to this type of information is probably too difficult an endeavour due to the sheer number of sites on the Internet where information on WMD technologies is present. The fact that much of this knowledge is dual-use in nature further complicates the determination of when information is being collected for WMD-related purposes. A less ambitious objective — assessing the distribution of information on global networks — and tracing the identities of groups collecting such information from these sites, might assist governments in their surveillance of potentially dangerous organisations and individuals. At the outset, however, the controversial nature of such surveillance over individuals who may be guilty of no criminal offense must be admitted. The surveillance of individual information gathering on the Internet is both technically difficult and politically and legally problematic. In the absence of a baseline appreciation of the information available, and of whom the consumers of this information are, it is difficult to see how investigations in this area can proceed.

Key Features of the Information Environment on Global Networks

Two features are fundamental to the workings of the Internet. First, the information available on the Internet is not under centralised control. This means that information of varying quality is available to almost anyone with network access. This also means that tracking the appearance and disappearance of new information repositories on the Internet is a very difficult enterprise. The second significant feature of the Internet is that information on almost any topic is now available to almost anyone. The quality of this information is subject to a wide degree of variation, but with skill it is undoubtedly possible to assemble very sensitive WMD-related information from public sources. The Internet adds speed and breadth to information retrieval efforts, however, increasing significantly the potential efficiency of information gathering by interested entities.

The combination of greater information access and increases in network connectivity creates a compound danger: new actor access to sensitive knowledge and materials important to the exploitation of WMD technologies, and a lucrative set of infrastructure targets where network-borne attacks can be launched against key societal assets. Verification in this environment means defining the distribution of sensitive information on global networks, characterising the "user base" of that information, and relating this information to other investigative resources used to combat proliferation worldwide.

To secure sensitive information networks from compromise, two things are necessary.

- Authentication systems must be deployed to ensure that only authorised users gain access to highly sensitive information resources. Such systems should probably be segregated from Internet connectivity in order to minimise the risks of remote penetration.
- Prior to the deployment of an authentication system an attempt should be made to construct a baseline measuring the actual level of intrusion attempts, the points of origin in cyberspace of their initiators, and the degree of success intruders display in penetrating secure systems. Current estimates vary widely, and constitute little more than educated guesses concerning the level of hacker 'probing' of secure information networks.

Securing information that is already freely disseminated within the Internet is, as was argued above, a rather futile endeavour. Maintenance of surveillance over the distribution of that information, and over the frequency and character of information collection by suspected proliferators could help to reinforce other anti- and non-proliferation measures taken by national governments and international agencies.

Characteristics of Internet-Available Information

Internet information resources comprise collections of data and structured indices of available data. Information resources are geographically separated, and access to this information is asynchronous — that is, the placement of information on a site is not related closely to the accessing of that information by a potentially anonymous actor in cyberspace. No centralised certification authority exists that establishes the veracity of particular information or analyses on the Internet.

Evaluating the distribution of WMD-related information available on global networks could be accomplished through the design of an intelligent software agent able to "roam" the Internet in search of information of a WMD-character. This type of agent would resemble existing Internet search engines, in that it would travel the global Internet, returning to the home-site with the URLs (Universal Resource Locator) of sites containing information on WMD-related topics. An automated tool of this type would need guidance from analysts capable of selecting among the vast number of URLs returned by the agent, with a view to re-targeting the search — and honing the intelligent agent's behaviour — to better characterise the nature of information available on the chosen topic.

Established traditions of openness and legitimate information access, alongside a developing notion of user anonymity, may inhibit the use of intelligent agent technologies to conduct verification activities on global networks. Three factors that exacerbate the difficulty of conducting these activities are:

- particular users of information may be difficult to identify, beyond the point of an institutional affiliation;
- information "copying" does not clearly deprive others of access, or reveal access activity if the creation of "mirror" sites is a frequent part of network activity; and,
- information may be derived from "processing" less sensitive network-derived data. This is particularly the case if skilled analysts are the investigators seeking the WMD-related information. It is possible that the pattern of information retrieval could be masked to conceal searches for sensitive information within more generalised searches.

If intelligent agent technologies offer the prospect of characterising the distribution of information resources on the Internet, the central domain of verification activities will most likely lie in the surveillance of accesses to known sources of sensitive information, and in the protection of content/information integrity. Three parts of the verification activity thus must be performed:

- surveillance of access to known sites of WMD-related information;
- protection of information integrity and security at government-controlled data centres; and,
- the construction of analytical frameworks that establish predictive models for the way suspected proliferators utilise information to make progress toward a WMD capability.

Categorising information can be performed in a variety of ways, but a simple example of a system for categorising information, and the entities seeking it, follows.

Categorising Information Gathering for WMD Technologies on Global Networks

If intelligent agent technology is applied to the task of evaluating the distribution of information resources relating to weapons of mass destruction on the Internet, a second task still remains, that of organising the information into a structure usable by analysts to interpret possible proliferator behaviour. It should be said at the outset that such a tool could only provide an incomplete picture of the information gathering activities of suspected proliferators. Global information networks are still new, which means that established information sources aiding proliferators exist outside of their parameters. It is likely that this situation will change as the information-rich character of global networks becomes more fully appreciated. In the interim, information on the Internet relating to WMD acquisitions must be categorised in a way that makes it usable to statistical and other analyses.

A Prospective Typology for WMD-Related Information

First, the category of weapon that a particular piece of information relates to can be noted. For the purposes of verification, WMD-related information could lie in one or more of four basic categories:

(1) technologies and materials relating to nuclear weapons;
(2) technologies and materials relating to biological weapons;
(3) technologies and materials relating to chemical weapons; and,
(4) technologies and materials relating to ballistic missiles.

In turn, the information can be further categorised according to type:

(1) dual-use technologies (system designs and subcomponents);
(2) proprietary datasets;
(3) test data from a weapon system; and,
(4) analytical reports and technology assessments.

Lastly, categorising the location of the sensitive information is also important, as this gives information regarding the places on global networks where information is most freely available:

(1) institution; and,
(2) country.

Categorised in this way, the information on the Internet can be analysed to establish the distribution of information available to potential proliferators. Tracing information accesses by individual actors could potentially be achieved through another intelligent software agent, or classifier system, that catalogued the Internet addresses of all accesses to particular Internet information resources over a measured period of time. Agent technology of this type is still largely experimental, but offers the potential to track activity that is already pre-screened through the use of other investigative tools. This second activity is necessarily controversial if used to track the Internet activities of ordinary citizens. Care would need to be taken to minimise abuses of network trace-back techniques.

Conclusions and Policy Implications

A brief summary of the features of cyberspace of importance to verification is followed by a listing of policy issues that require consideration in the design of responses to potential problems.

Features of Cyberspace

- Cyberspace can be seen as a channel for the transmission of proscribed information relating to military technology — especially that concerning weapons of mass destruction. International treaty or arrangement — i.e., NPT, CWC, BTWC, AG, MTCR, and the Wassenaar Arrangement — proscribes much of this information. The unprecedented access to dangerous information through cyberspace is now potentially available to a wider range of actors beyond traditional government agencies, including a host of non-state actors.
- Cyberspace can be seen as a target or environment for new kinds of weapons and warfare — information war. There is a proliferation of 'new weaponry' and new vulnerabilities, which might conceivably be the subject of new non-proliferation, arms control, and disarmament measures that attempt to address these new potential threats to national and international security. Developing negotiations on responses to still hypothetical dangers may be difficult, especially given the other issues already on the international arms control agenda. A poor understanding of these issues may also impede rapid policy responses.
- Cyberspace can be seen as a potential new verification tool, or, perhaps more accurately, a new environment for using tools to verify compliance with extant state obligations relating to non-proliferation of weapons of mass destruction. For example, cyberspace monitoring methods may offer new information regarding the activities of potential proliferators around the world.

From these insights flow a number of implications:

- Non-state actors are likely to figure prominently in the entities that exploit the new information warfare capabilities for disruptive or destructive purposes. Economic, political, and religious stakes are the subjects likely to generate proliferation-related behaviour.
- Non-state actors gain disproportionately from the spread of information networks to new areas of the world. The opportunity for new coalitions of NGOs to share information and mutual support offers social forces an impressive new resource for influencing government policies. In turn,

new networks offer new capabilities for actors with network access to disrupt key information systems for political or criminal purposes.

- The pace of technological change may channel proliferator behaviour in directions where few developed taboos exist against the use of disruptive weapons with widespread effects. This potential, alongside the progressive empowerment of non-state groups relative to government authorities in cyberspace, may create dangers as non-state actors use disruptive information weapons and perception management to pressure governments to change policies.

Identifying Critical Infrastructures

Thus far, the danger to critical societal infrastructures such as banking, electric power, and telecommunications applications are entirely hypothetical — in that no attacks have occurred or, at least, been detected. The critical nature of these infrastructures is not in doubt, however. Nor is their increasing use of open system architectures utilising network and Internet-related technical standards. The adoption of these systems and control protocols means that weaknesses and security shortcomings impacting the 'known' Internet are likely to propagate inside increasingly interconnected wide-area information networks. Investments in network security in each of these infrastructures offer the best near-term prospects for avoiding increased abuse of networked information systems. Prior to policy initiatives in this area, however, risk assessments and vulnerability analyses should be performed to evaluate information-dependent infrastructures. The following recommendations point in this direction.

- An analysis should be undertaken of the scope and nature of critical infrastructures adopting Internet-linked remote operation and maintenance applications in Canada. In the United States, an infrastructure protection task force has been established to address some of these issues. A similar effort should be launched in Canada, with a view to assessing the vulnerabilities of key infrastructures to compromise.
- A comprehensive survey should be undertaken of the sources on the Internet which provide weapons-related information in all areas of WMD concern. In addition, consideration should be given to applying emerging intelligent agent technology in order to characterise the information available on the global Internet in terms of both its importance and geographic location. Research could then begin to consider measures necessary to ensure that proliferator exploitation of these information resources is under the appropriate high level of surveillance.
- While consultations on infrastructure security between countries are probably premature, efforts should be made to collate the findings of the many parallel national efforts at assessing the implications of the growth of network connectivity for infrastructure and national security. Independent efforts in this regard have been completed in the United Kingdom

and the United States, but no integrative study of the cross-national growth of infrastructures — and of the parallel growth of network vulnerabilities to disruption — has thus far been attempted.

- Because economic actors are driving the construction of cyberspace, national security and regulatory concerns are frequently forced to 'catch up' to the changing shape of a rapidly evolving technology. To come to terms with this situation, a forward looking study aimed at projecting the future actors likely to take advantage of network connectivity for nefarious purposes should be considered. The actor-goal matrix presented in this chapter suggests that extortion and intimidation are likely to dominate the use of information weapons by non-state actors. This reality poses new challenges for governments in that threats to national security — interpreted as threats to critical infrastructure security — may resemble terrorist or criminal activity rather than traditional acts or threats of war.

While states may eventually achieve agreement on the sanctity of information infrastructures from attack, non-state actors may refuse to respect developing taboos. This means that states may confront a threat that is difficult to detect or fully evaluate. The danger also exists that suspected proliferators will use non-state actors as 'fronts,' concealing the true author of particular cyberspace exploitation schemes. Establishing the defining features of the new network environment is a critical first step to responding appropriately to the new information challenges present there. Only then will the truth regarding emerging societal vulnerability to network attack be fully appreciated.

Chapter Eight

The Information Revolution, the Military, and Arms Control

Jeffrey R. Cooper and Christopher Burton*

The "Information Revolution" is without a doubt one of the most talked about subjects of the 1990s. Newspapers, magazines, and television commentators ceaselessly discuss its technologies, its progress, and the consequences it will presumably have for every facet of life on earth. Within this bewildering 'universe' of subject matter, this chapter will attempt to examine only a component 'galaxy': it will discuss the implications of these new technologies and new ways of procuring technology for the military, and, by extension, for arms control. To do this, the chapter will proceed through four sections. First, it will position itself within the many subjects related to the Information Revolution, to make it clear which of the many implications will be further discussed. Second, it will describe how information technology is being used by the military, and how technological and economic changes resulting from the Information Revolution have affected how the military procures equipment. Third, it will describe at greater length the technological developments that together make up the Information Revolution, in order to provide a common basis of knowledge upon which to build discussion. Finally, based on this summary of the technologies of concern and how the military uses them, the chapter will conclude by discussing some of the implications of the Information Revolution, both in the broader realm of national security and in the area of arms control efforts in particular.

Broad Implications of the Information Revolution

Whatever the specific technological drivers for the Information Revolution, we can now clearly discern some of its key implications for both the military and civilian sectors in information processing, command, control, and communications. One of these impacts will dominate all others — the intermingling of civilian and military technologies and users to the point where they will be impossible to disentangle. This impact will demand dramatic changes

---

* This chapter is an edited version of a presentation made to the Fourteenth Annual Ottawa NACD Verification Symposium, "Cyberspace and Outer Space: Transitional Challenges for Multilateral Verification in the 21st Century," 12–15 March 1997, Montebello, Quebec, Canada.

in how we think about potential controls over the diffusion of these new information technologies.

Beyond this overriding impact, there are a number of other potential implications that deserve specific mention. First, this revolution presages a flood of information, previously unavailable, that now will be widely disseminated near-instantaneously to interested users located anywhere — and much of this information is derived from a variety of advanced sensors that are owned and operated by civil or commercial entities. Second, this information will be supplied in quantities and at speeds that were previously unattainable, yielding completely new functionality in its use. Masses of data previously unusable or uninteresting in their discrete, atomistic form will be merged, correlated, and interpreted into useful information and knowledge. These new capabilities will raise significant economic, legal, and social issues about privacy, data security and integrity, intellectual property rights, and other appropriable benefits resulting from these activities. Third, this revolution provides significantly increased utility from even previously available processed information by synthesizing, integrating, customizing, indexing, and compressing it — changing the nature and role of data storage and archives, libraries and encyclopedias, movies and videos. Fourth, this revolution in communications technologies gives capabilities to transmit these vast volumes of information essentially on demand, with high assurance and low costs. This degree of portability and mobility for information and information users is unprecedented. Fifth, together, these capabilities allow coordination of diverse individuals and activities in ways never before possible. How these "virtual" organisations, communities, and teams will alter existing work processes cannot yet be determined. Finally, real-time encryption provides privacy and security never before available. But, at the same time, concern over vulnerability of our information systems, and about those systems dependent upon them, has never been higher.

Information Technology and the Military

For the American military, and for militaries throughout the world, the emergence of the Information Revolution portends a range of changes that will eventually reshape underlying assumptions and transform not only military operations — including both their character and execution — but military organisations as well. Throughout the Cold War, the U.S. military believed that its information, command and control, and communications needs were exceptional, if not unique, and unmatched in the civilian or commercial worlds — views probably held by most other advanced militaries. Therefore, it believed its needs demanded the most advanced sensors, the most real-time data collection capabilities, the most advanced computers, and the widest bandwidths, thereby allowing the military to set the nation's R&D agenda and control the rate of technological change, at least in this crucial arena. However, while military users may continue to have *particular* needs, they have now lost the vanguard position so recently held, because these demands are no longer exceptional, either in quality or timing.

At the most obvious level, the Information Revolution offers militaries potentially dramatic improvements in force effectiveness from three types of potential applications. First, as foretold during the 1991 Gulf War, seamless "sensor-to-shooter" links offer the ability not just for 'smart' but 'brilliant' weapons launched from survivable stand-off ranges. Second, these same technologies can provide military commanders and decision-makers with real-time information and insights enabled by Dominant Battlespace Awareness. Finally, these technologies can allow seamless real-time integration of air, sea, and land forces in a common tactical operating environment.

Less noticeable, four key structural developments wrought by the Information Revolution upon the intelligence and communications environment, to which the military had become long-accustomed, also demand acknowledgment:

- the opening of the military's traditionally closed, autarkic C^3I environment;
- the military's loss of technological leadership to the commercial sector;
- the reversal of many of the long-standing technical and cost relationships that prevailed in the military sphere; and,
- the military's need to accommodate itself to the commercial sector's pace of change.

Opening the Closed Communications Society. For most of the post-Second World War period in the communications domain, the military constituted a "closed," autarkic society. This was appropriate for the particular circumstances of the nuclear confrontation between the United States and the Soviet Union. However, since then, the military has moved rapidly from the highly secure, closed system that even five years before it controlled end-to-end — from R&D through acquisition to operational use — to today's increasingly open system. In this new environment, core military C^4I functions are deeply embedded, if not inextricably entangled, within a web stretching from the Defense Information Infrastructure, through the National Information Infrastructure, to the global information infrastructure. Today, fully 95 per cent of day-to-day U.S. defence communications are carried by commercial channels, mostly through the public switched network.

Losing Technological Leadership. Even if the dramatic reductions in defence expenditures since the depths of the Cold War had not occurred, the civilian commercial sector might still have taken the lead in developing many of the new technologies as the opportunities for economic efficiency and profit became evident after limited commercial exploitation. Esoteric technologies that previously appeared to have few civilian applications are now being rapidly applied across a wide range of civilian activities, both for commercial and domestic purposes. These include widespread adoption of advanced telecommunications systems on a global basis, incorporating the new digital technologies in everyday appliances, application of GPS-based navigation systems to personal use, and employing extremely advanced sensors to support real-time production and control systems. These changes have substantially altered the direction of technological development in computers and semiconductors — for example, Silicon Graphics, Inc. now lists commercial customers, especially in the entertainment industry, as its key "Lighthouse Partners" for defining leading-edge technology requirements.

Reversal of Key Relationships. The military must now look toward procurement of commercial off-the-shelf, dual-use equipment, and open systems architectures, instead of relying on the classical but autarkic military R&D and acquisition systems which produced highly optimized and cutting-edge products, but at high-cost and usually with proprietary technologies. Hardware is now cheap while special purpose proprietary software is costly, but maintenance of "legacy" systems — both hardware and software — is now prohibitively expensive.

Rapid Pace of the Commercial Sector. The military has found itself spurred into adopting more rapid acquisition processes, both to keep up with a rate of technological advance set by the commercial sector employing a common digital technology base that the military no longer dominates or drives, and to reduce the costs of its outdated, autarkic legacy system. The proliferation of new telecommunications services, such as data, cellular, and personal communications systems (PCS), has created many new companies, which not only serve these new markets, but also compete with existing services. This competition, in turn, further spurs the development of advanced, competitive-advantage producing technologies.

Together, these changes amount to the military becoming only one among many users and, in most cases, not even *primus inter pares* — a purchaser of systems rather than a developer, a follower of technology rather than an advancer of the state-of-the-art, and a buyer of services rather than an operator. The shift from a national security-dominated to a commercially-dominated information environment means that the national security community will no longer drive the rate and direction of innovative activity, the types of systems produced, nor the establishment of formal or *de facto* standards. These changes will demand significant cultural adjustments, not only by militaries but by the arms control community as well.

A Constellation of Advanced Information Technologies

The technological component of what we call the 'Information Revolution' results from a set of related technical innovations, together yielding substantial enhancements and perhaps even the revolutionary change that the term itself promises. What we are witnessing does not appear to be a case of inexorable 'technology creep' but rather an explosive 'technology rush.'

Every two years, the speed of silicon processors doubles and storage costs decline by half; every five years, compression efficiency is squeezing data to 1/30 of its size; and every three and a half years, laser diode speeds, critical for fiber optics, are doubling. Terabit transmission speeds have already been demonstrated in the laboratory.

These capabilities support a full panoply of advanced sensors that gather more and different data, new communications architectures — including "untethered" mobile systems — that change fundamental cost relationships for information storage and distribution, and mobile "supercomputer" capabilities married with seamless integration that multiplies individual component functionality and effectiveness — all based on continued progress in digital microelectronic technologies.

The Digital Revolution

Perhaps the dominant contributor to this constellation of new technological capabilities is digital technology itself. With the accelerated introduction of computer and semiconductor technologies into almost all facets of our existence, there is increasing convergence of commercial telecommunications, entertainment, home electronics, and computers.

Digital technologies give rise to two key implications that require significant changes in the information user's frame of reference. First, information in digital form is replicable at will, without degradation or change from generation to generation. Second, digital data streams are inherently encrypted by the process of converting data into streams of binary digits. Because real-time algorithmic manipulation is possible, extremely high security coding schemes are feasible at low cost. Indeed, they allow real-time on-line encryption that allows secure *synchronous* communications. Moreover, these capabilities can be applied not only to written or data communications, but to voice and video as well.

At the same time, these digital technologies are raising fundamental questions concerning ownership and control of intellectual property, privacy and security of valuable data, and equities contested between law enforcement and national security communities on the one hand and commercial interests on the other. Not only for the military, but for the national security community as a whole, these complex changes must be assimilated even as these organisations continue to adjust to the significant structural changes produced by the end of the Cold War.

Ubiquitous Computing

The second key contributor to the Information Revolution is the continuing advance of computing power. Computers are now 100,000 times faster and 1,000 times less costly than during the 1950s, with even greater increases in functionality due to graphical user interfaces, modern software, networking, and platform integration. There is little indication that Moore's Law, predicting doubling of memory chip capacity every eighteen months, is 'slacking off' very much from its historic rates of advance — it will probably continue for at least the next decade. The rate of increase in computational speed, in fact, appears to be speeding up, with recent improvements averaging about 55 per cent per year, compared with 35 per cent in 1985. Current computational power has outpaced previous predictions made in the early 1980s by a factor of more than 100.

Four technological innovations are components of this remarkable increase in computing power.

"Supercomputers" for Everyone. The use of cutting-edge machines which, just a few years ago, were available only to select military and other national security users, is now firmly established in everyday commercial use and best practices.

Embedded Computers. The same solid-state technologies — semiconductor microprocessors and memories — that allow desktop-sized supercomputers also enable the computational heart of these instruments to be embedded in other machines and systems, thereby fundamentally altering the effectiveness and usability of commonly available tools, appliances, and even complex systems.

Cheap, Random Access Mass Storage. A critical complementary technological breakthrough is the availability of high-capacity, random access mass storage, now measured in gigabytes and terabytes of instantaneously accessible data. Portable archives of this size allow complete map and geographic databases to be locally resident, requiring only change functions to bring them instantaneously up to date.

"Highly Cognitive Displays" and User-Configurable Interfaces. Perhaps the next frontier that technological advances in semiconductors, computation, storage, and software will address is effective processing, individuation, and display of complex information in order to allow it to be easily and correctly assimilated by the user. Increasingly large-scale integration of networks and synthesis of useful information from massive, complex data archives for commercial customers is providing the driving force for advanced representation and display systems. Virtual reality technology is creating new ways of displaying information other than as static data sets.

"Infinite" Bandwidth

A third key element in the Information Revolution is cheap, near-infinite bandwidth transmission capacity that creates an entirely new communications environment. Progress in this area is largely the fruit of continued advances in the same solid-state digital technologies that have increased computational and storage capabilities. Here, embodied as signal processors, amplifiers, transmitters, and receivers, these advances have continued to press usable frequencies ever higher. K_a-Band (20–36 GHz) satellite transmission capability and similarly wide-band capacities in advanced fiber-optic cables represent fundamental advances in information transmission capabilities that do not appear to be slowing down. Indeed, as mentioned above, recent laboratory work has experimentally demonstrated terabit transmission capacity using advanced laser diodes and optical fibers. In creating these systems to meet civil and personal needs, the commercial sector is constructing a global information infrastructure that meets many of the military's demands, and in some cases, meets latent needs that have not yet even been recognised.

"Untethered" Communications

A fourth area of innovation that has contributed to the technological revolution of the present day is the advent of "untethered" communications. Historically, there were few users, other than the military community and steamship companies, who expressed a need for mobile, as opposed to relocatable communications capabilities — either to transmit orders to moving units whose location was uncertain, or to receive situation reports from those units. However, the unbelievable speed with which civil society throughout the world has adopted cellular telecommunications and now PCS services testifies to a widespread latent demand for "anywhere, anytime" communications.

In 1995, wireless communications generated more than $22 billion in revenues and this growth, with 20 million domestic users, outstripped predictions made less than a decade before by a factor of more than 100. By 2001, it is forecast that, worldwide, 500 million people will use mobile communicators such as cellular or PCS — whether linked through terrestrial or satellite networks. Furthermore, radio modems will link laptops and other mobile information processors and disseminators back into the entire national or global information infrastructure. To meet these demands, at least seven major space-based mobile-user communications systems are now planned, covering a variety of service features and price points.

Sophisticated Sensors and Discrimination Algorithms

The only point that must be made here is that many critical military needs can now be met by commercially available remote sensing systems. Space-based sensing, once the sole province of the two superpowers' military establishments, is rapidly becoming an essential adjunct to day-to-day civil society, and not only for developed nations. Sensors increase the reach, decrease the time for measurement and perception, and extend the spectrum well beyond direct human senses and experience — they can make real what before was not even perceived.

Civil society has discovered the value of increasingly sophisticated information-related capabilities and commercial entities are inventing ways to turn these demands into profits by employing the most advanced sensors for data collection — previously only available to the national security community — across an extraordinarily wide range of applications. The Russians are currently selling better than one meter resolution — formerly secret compartmented data from reconnaissance satellites — and by the end of next year, a commercial American satellite will be providing similar quality data. For many civil uses, this high resolution data is essential. It has become increasingly difficult to make distinctions between civilian and military requirements and satellites.

For example, formerly secret data and classified techniques and systems are in high demand by oceanographers and oil companies. Civil or commercial systems employing extraordinarily sophisticated sensors include advanced air traffic control systems, Doppler weather radars for wind shear and storm warnings, space-based environmental monitoring systems, and supervisory control and data acquisition (SCADA) systems for monitoring and control of large-scale telecommunications, energy transmission, and continental and regional transportation networks that affect millions of people in their daily lives.

Two uses in particular stand out:

Precise Navigation. The Global Positioning System (GPS) was developed by the U.S. military to allow military units to determine their location with precision by comparing the time taken by radio signals from many different satellites to reach them, but the civilian GPS industry and the number of civilian users now overshadow the military.

Mapping and Monitoring. The rapid growth of sophisticated geographic information systems enabled by high resolution maps is resulting in a wide range of civil and commercial applications, including crop management, store location, flood control, and emergency response.

Implications of Advanced Information Technologies

Implications for the Military and National Security

These various technological developments, when considered together, do indeed amount to a revolution — the effects of which will be felt in almost all areas. Perhaps the greatest effect overall of the new information technologies is the breaking down, across many dimensions, of traditional distinctions and boundaries between different information-related elements — including those that distinguished and separated the military and civilian sectors in their demands and uses of advanced information technologies. This convergence is taking place in a wide array of different spheres:

- in the technological dimension, underwritten by a common technical base of solid-state digital microelectronics;
- in the user dimension, as individuals no longer want to distinguish among information sources and providers, nor employ different instruments to tap them;

- at the system level, where network architectures increasingly must be prepared to supply seamlessly integrated multimedia products;
- in the geographical dimension, as physical and political boundaries, time and space, lose their classical divisive impacts;
- in the national security dimension, as the traditional distinction between raw information and processed intelligence disappears due to both the widespread availability of relevant open source information and the demands of operationally-oriented users for usable real-time actionable knowledge; and,
- in the legal and regulatory communities, as the old distinctions and barriers between different information forms and regimes collapse.

Three overall consequences flow directly from this revolution in information technologies, capabilities, and employment. The first is that the widespread adoption of these merged, cutting-edge information technologies has created a vast commercial and consumer constituency that overshadows the national security community, in the U.S. and elsewhere. As a result, the previous priority and obeisance accorded to national security interests is not likely to continue. Moreover, a host of controversies that would previously have been settled on terms favourable to the national security community can now be expected to be decided in favour of commercial and civilian interests.

As a second important consequence, these new information and telecommunications capabilities are creating a truly global information infrastructure and environment in which traditional notions of geography and spatial separation no longer exist. Geographic distance no longer implies time delays in communication between any two points. The increasingly intertwined nature of the "infosphere" also means that there are no impenetrable boundaries between domestic domains or even nations, and this loss of geographic reality has exposed the formerly "safe" rear areas of nations' domestic infrastructures. Thus, there are no sanctuaries; even formerly secure rear areas are now as exposed as front-line forces to information attacks by opponents. Moreover, distinctions between military and commercial functions are increasingly meaningless when over 95 per cent of defence messages move over the public switched network, when just-in-time logistics for forces engaged in combat 'reach back' into contractor facilities, and when commercial contractors are providing real-time communications and analysis services in support of ongoing military operations.

A third important consequence of our Faustian acceptance of the 'Information Revolution' bargain is the emergence of a collective national infrastructure — electricity, telecommunications, oil and natural gas, freight, air traffic control, and so on — that is increasingly reliant on real-time scheduling and processes and is fully integrated with a national information infrastructure that is, in turn, increasingly integrated with a global information infrastructure. These systems are then in their turn often monitored and controlled remotely through SCADA systems that are themselves networked telecommunications systems tied into the same potentially vulnerable infrastructures. These trends are producing a system that is increasingly efficient, but is stable only in its dynamic configuration and may not degrade gracefully.

Another set of consequences flowing from the Information Revolution will also have an effect on the military and national security. However, these will have their effect through the intervening medium of information security. For several key reasons, the technological changes of the Information Revolution portend significant implications for effective concepts of information security and protection. First, the widespread adoption by both military and civilian/commercial users of common, commercial off-the-shelf equipment, systems, and practices decreases the difficulty that potential hackers or attackers will have in identifying entry points or finding vulnerabilities. At the same time, military communications carried over shared dual-use systems or in common frequency regimes with standardised protocols may find reduced vulnerability, both from being "lost in the noise" of a massive civilian/commercial traffic volume and from increased resiliency due to massive redundancy resulting from the complex network of alternate service providers in the commercial domain.

Second, while the military and intelligence communities have traditionally been more concerned with system security and survivability than commercial users, the development of new commercial and civil uses that are real-time and "mission critical" means that commercial users may be just as concerned in the future about privacy and security. Widespread adoption of real-time, on-line transaction services and SCADA systems for critical infrastructure systems creates a fundamentally new demand for security, privacy, and verifiability, especially where data flows involve billions of dollars or can put millions of lives at risk.

Third, widespread access by news media and others to advanced real-time sensor and communications systems may create an entirely different set of security and information protection concerns. With increasingly free access up and worldwide direct satellite broadcast down, television and radio news coverage will largely be outside of government control over content and timing. The availability of small, direct-uplink, high-capacity systems capable of supporting real-time video together with demands from news organisations for real-time information, unfiltered or uncensored by governments, is likely to produce substantial discomfort as security interests clash with claims of press freedom.

Implications for Arms Control Efforts

Just as these technologies have important consequences for the military and the practice and nature of war, so too do they have implications for arms control and disarmament efforts and regimes. The implications, however, are double-edged. First, new information technologies could prove to be important new tools in the cause of arms control, because of their potential uses for verification and information gathering. As powerful technologies spread beyond the armories of the mightiest nations into the commercial sector they also become available to the international arms control community. If sensor data can be bought, it can be bought for the purposes of arms control verification. These technologies increase information and transparency, and from an arms control perspective, that must, on balance, be good.

On the other hand, these new technologies also pose arms control challenges. The same cheap and powerful computing, communications and sensor capabilities that are revolutionising the global economy also put a whole new magnitude of military power within the reach of whomever has the knowledge and tools to take advantage of it. However, attempting to create a new arms control regime to contain these innovations — to impose old-style controls in order to limit diffusion of potentially dangerous technologies — is not likely either to be feasible or effective. Controls on railroads in turn-of-the-century Europe would have been unthinkable, despite the crucial role of railroads in the military mobilisation plans of the time — however important the railroad's military uses, its civilian uses were of even greater importance. In the same way, restrictions upon information technologies with a view to limiting their military uses would of necessity limit their civilian uses as well.

These technologies are fast becoming the eyes, ears, and nerves of our society, and any concerted attempt to limit their spread would therefore be self-destructive as well as politically impossible. Today, mass market technologies such as desktop computers and GPS devices theoretically pose proliferation threats, but controlling such basic products would be impossible without severe economic and political repercussions. Furthermore, technological change now occurs over a cycle measured in months — far too fast for any legislative attempt at arms control to keep up with. In truth, any attempt at controlling the proliferation of information technology would be starkly irrelevant to the environment of today where the technological cutting edge moves 'in the blink of an eye.' It would therefore merely serve to imply progress where there was none and so disguise the futility of the underlying approach. In the end, we are better off with a real sense of vulnerability as the Faustian trade-off for technological capability than with a false sense of security bought at the price of economic progress — this, at least, should keep us on our toes.

Chapter Nine

Virtual Security: Technical Oversight, Simulated Foresight, and Political Blindspots in the Infosphere

James Der Derian*

> To be means to be for the other, and through him, for oneself. Man has no internal sovereign territory; he is all and always on the boundary; looking within himself, he looks *in the eyes of the other* or *through the eyes of the other* ... I cannot do without the other; I cannot become myself without the other; I must find myself in the other, finding the other in me.
>
> Mikhail Bakhtin, *The Problems of Dostoevsky's Poetics*

> Alienation is no more: the Other as gaze, the Other as mirror, the Other as opacity — all are gone. Henceforward it is the transparency of others that represents absolute danger. Without the Other as mirror, as reflecting surface, consciousness of self is threatened with irradiation in the void. ... No longer the hell of other people, but the hell of the Same.
>
> Jean Baudrillard, *The Transparency of Evil*

In an escalating order of concern, four questions inform this chapter. How does one approach a phenomenon so ubiquitous yet so elusive as surveillance? How does one theorise — which from its Delphic origins (*thea* and *horao*) means 'to attentively look outward at something' — a technology that looks back at the theoriser with the reflected arrogance of science, a gaze that offers global knowledge dressed in the guise of objectivity and transparency? How does one criticise something that has been deemed vital not only to national security but also to corporate, environmental, family, and personal security? How does one offer a plausible alternative to the collective belief that we live in a world at risk, and that our ability to foresee, perhaps even to forestall,

---

* This chapter is an edited version of a presentation made to the Fourteenth Annual Ottawa NACD Verification Symposium, "Cyberspace and Outer Space: Transitional Challenges for Multilateral Verification in the 21st Century," 12–15 March 1997, Montebello, Quebec, Canada. A version of this paper was presented at the 38th Annual Convention of the International Studies Association, "Coping With Insecurity: Threats More Than Enemies," Toronto, Ontario, Canada, 18–22 March 1997.

danger requires a technology of surveillance which can oversee everything and everybody?

There are some extant theoretical responses, but each comes with shortcomings. Modernism, wedded to the idea of progress through technology, is deeply implicated by surveillance in the workplace, at home, on the battlefield, and indeed, by the mimesis of positivist modeling itself [1]: it is hardly conducive to the kind of intellectual distancing needed for a critical inquiry. Pre-modernist approaches offer historical depth and narrative breadth, but cannot explain, let alone anticipate, the structural effects of rapid changes caused by innovations in surveillance techniques. Post-modernist approaches, like critical genealogies and intertextual analysis offer a deeper sensitivity for the de-territorialised, chrono-political, and global effects of surveillance, but often fall short in the area of policy alternatives.

This array of practical and political difficulties prompts many academics to take the high road of meta-theory, to theorise about theory, or, in my case, to put surveillance under surveillance. Aside from a few conceits offered as meta-theory, this will not be the strategy of this chapter — I've been there, done that.[2] Besides, I have come under other more powerful influences. I have been compelled to throw caution, commensurability, and the cloak of meta-theory to the winds, and to morph pre-, post-, and modernist approaches for this investigation into surveillance. I could provide a host of intellectual justifications, but the reason why has become too difficult to conceal: too many viewings of *The X-Files*. The programme's ('pre'- 'post'-erous) slogans have become my epistemological mantra: 'The truth is out there' (often *way* out there), 'Trust no one' (*especially* the truth-sayers), and 'The Government denies knowledge' (an acknowledgment of guilt, ignorance, *and* epistemic drift). Radical measures, perhaps, but after a year in which a presidential candidate exalts *Independence Day* for its American values (a movie in which a replica of the White House is bombed by aliens),[3] the FBI's surveillance profiling turns a security guard at the Summer Olympics bombing 'from national hero to public zero' (a *Daily News*' headline from 15 August 1996)[4],

1. See Richard Ashley, "The eye of power: the politics of world modeling," *International Organization* 37:3 (Summer 1983), 495–535.
2. See James Der Derian, *Antidiplomacy: Spies, Terror, Speed, and War* (Oxford: Blackwell, 1992), 19- 39; and "Antidiplomacy, Intelligence Theory, and Surveillance Practice," *Intelligence and National Security Journal* (July 1993).
3. *Independence Day* also made the 8 July covers of *Time* ("Aliens have landed!") and *Newsweek* ("'Out there' — from *Independence Day* to *The X-Files*, America is Hooked on the Paranormal"). And as the corner ribbon: "Terror in the Gulf: How Shaky Are the Saudis?"
4. "From national hero to public zero," *Daily News* (15 August 1996). When asked whether there is a real *X-Files* division at the FBI, former supervisory special agent Gregg McCrary replied: "As close as we come is the unit I was in, the 'Profiling Unit.' It deals with the behavioural sciences, and bizarre and unusual crimes — sex crimes,

and International Relations (IR) continues to confront increasingly *irreal* events, there is an even greater need for one's theoretical reach to exceed a discipline's grasp (or what's the extraterrestrial for?).

So I am a confessed *X-File*-phile. In fact, the origin of this chapter stems from an invitation for a conference on surveillance in Vancouver, the very city where *The X-Files* is taped — an invitation that I could not refuse. When I returned to Canada to attend the Annual Meeting of the International Studies Association (ISA), those tireless purveyors of truth, FBI agents Scully and Mulder, inspired new investigations. Flashing my ISA badge to gain entry into panels, plenaries, and publishers' booths, I have discovered that alien forms have infiltrated into nearly all ranks of IR theory. They have come as green-blooded, shape-shifters (a.k.a., the constructivist cabal), resistance-is-futile borgs (the critical security studies thuggees), gender-bending droids (the feminist coven), and, of course, those big-headed, bug-eyed mutants (the post-structuralist conspiracy).

This is not good news to the Syndicate: on *The X-Files*, they are those 'old white guys' who run the whole show from a smoky room somewhere on the east coast of the U.S.; in International Relations — well, just scrub the smoke from the picture. The job of the Syndicate, as one tells Scully, "is to predict the future, and the best way to predict the future is to invent it"; not too far removed from the self-fulfilling prophecies of the 'neo(realist)-neo(liberal)' Synthesis in IR.[5] But the Syndicate has developed a peculiar relationship with the aliens. Unable to destroy what they fear, they now seek to control the aliens, first by suborning the shape-shifters — one of whom, the Bounty Hunter, becomes employed as a very scary terminator — and then by setting up a big research programme called 'Purity Control,' whose goal is to extract DNA and other vital elements from the bug-eyed aliens for the production of hybrids (check your ISA programmes for details). This season opened with an episode in which Mulder was led to a Canadian farm tended by a cult-like community of clones. All of them were dead-ringers for his sister, Samantha, who, at eight, was abducted by aliens (or was it actually by the Syndicate?). When asked what the clones were seeking to achieve, she replies, "Hegemony." By the March sweeps, however, it is no longer clear whether the Syndicate controls the aliens, or the aliens had gained control of the Syndicate. Had the aliens come to save Earth from its own kind? Extraterrestrial eco-tourists on a mission of mercy? Or had they become pawns of the Syndicate — a kind of post-war solution for the fall of the 'evil empire'? Worse, was

---

serial killers, etc. There isn't any real need for an *X-Files* because there aren't any aliens running around — but there's enough bizarre human behaviour to keep us busy." "The War of the Worlds: In an Alien Nation, Whom Do You Trust?" *Entertainment Weekly* (29 November 1996), 35.

5. See Ole Waever, "Figures of international thought: introducing persons instead of paradigms," *The Future of International Relations: Masters in the Making?* Iver B. Neumann and Ole Waever, eds. (London and New York: Routledge, 1997), 1–37.

there actually no one in control, and everyone afraid to admit it? Stay tuned for the next ISA meeting.

Aliens aside, let me get my own meta-theoretical conceits out of the way. Surveillance is Heaven (God). Surveillance is Hell (Sartre). Surveillance is fetishised desire (Freud). Surveillance is herd resentment (Nietzsche). Surveillance is patriarchal (Lacan). Surveillance is good (the V-chip). Surveillance is bad (the Clipper chip). Surveillance disciplines (Foucault), dissuades (Virilio), simulates (Baudrillard). Surveillance is everywhere (Agent Mulder). Surveillance is in your head (Agent Scully). My sole philosophical aim, to paraphrase Gilles Delueze (and to put in a plug for the new on-line journal, *Theory & Event*),[6] is to make theory worthy of the event, not by determining the cause of an event, but by interpreting the powerful and often ambiguous effects of surveillance.

This means that this chapter comes uncomfortably close to the kind of media spasm that reduces all phenomena and events — revolutionary or not — to superficial and ephemeral forms. Probably the greatest challenge in an age of information revolution is to slow down, to down-shift from media-hype and fast-and-easy stereotypes, to down-play crisis mongering and crisis-management for more deliberative and experiential forms of analysis and decision-making. But this is not my immediate intention. This chapter moves from screen to screen, montage to holograms, sound-bites to buzz cuts, from substance to style — for nothing is so powerfully insubstantial, multiply mediated, and simulated as information — to alert us to the dangers rather than to pretend a solution for the most profound effects of the information revolution.

The bulk of my book *Antidiplomacy* is taken up with the theme that the current effects of surveillance cannot be isolated from the effects of simulation, and that they are more profoundly produced and sustained by an acceleration of pace rather than an occupation of space. My goal is to show how simulations through surveillance, from radar gun to spy satellite to computer screen, works as a technology and works on us as a technique of power through its ability to oversee and foresee, speed-up and slow down flows of information, capital, troops, refugees, drugs, viruses, and pollutants. In short, not the deep identities of geopolitics but the transparent differences of chronopolitics, where power is more 'real' in time than space, it comes from an exchange of signs rather than goods, and it is transparent and diffuse rather than material and discrete.[7]

---

6. http://muse.jhu.edu./journals/theory_&_event

7. See James Der Derian, *Antidiplomacy: Spies, Terror, Speed, and War* (Oxford: Blackwell, 1992); and James Der Derian, "The (S)pace of International Relations: Simulation, Surveillance, and Speed," *International Studies Quarterly* 34:3 (September 1990), 295–310.

Alloyed by the always renewable threat of terrorism, surveillance, simulation, and speed form the undertheorised, overdetermined currency of the information revolution, and as such, it is inseparable from the issue of security. National security is endangered by too little (or too much) information; computer security is necessary to prevent the theft or invirillation of information; the Securities and Exchange Commission draws a line between inside and outside information, and secures the borders of high capitalism; and at conferences like the ISA's we negotiate various meanings of security through our sharing (persuasion), withholding (manipulation), or distortion (propaganda) of information. We can freeze-frame factors of surveillance, simulation, and speed in all of these forms of security. But my brief is to focus on one particular aspect, the new surveillance effects of the so-called 'Information Revolution.' Again, this is my attempt to get at the truth of the matter while trusting no single version.

Cut to CNN, General Shalikashvili, testifying before the Senate Foreign Relations Committee on the bombing in Saudi Arabia, 9 July 1996: "Terrorism will always take the most indirect approach."

Follow with Paul Verhoeven, director of *RoboCop, Total Recall,* and (regretfully) *Showgirls*: "The U.S. is desperately in search of an enemy. The communists were the enemy, and the Nazi's before them, but now that wonderful enemy everyone can fight has been lost. Alien sci-fi films give us a terrifying enemy that's politically correct. They're bad. They're evil, and they're not even human."

Buzzcut back to Captain Kelvin Davis inside Kuwait City: "I hate to say it, but once we got rolling it was like a training exercise with live people running around. Our training exercises are a lot harder."[8]

Buzzcut forward to now (mid-March 1997), and the National Training Center (NTC) in the high Mojave Desert, where Fort Hood's 1st Brigade of the 4th Infantry Division, kitted out with $250 million in computers, satellites, and digital links, is about to face the Army's OPFOR ('Opposition force') for the fourth digitised rotation at the NTC. From *USA Today Online*, "Cybersoldiers test weapons of high-tech war" (6 March 1997):

> The home team will try to find the Achilles' heel of the new system. On recent maneuvers, the high-tech suppliers swarming over the field headquarters, the Tactical Operations Center (TOC), make it look like a movie set with the soldiers as actors. Five dozen new devices are being prepared for action. The soldiers are making progress. In less than six hours, the sweaty troops have transformed an empty clearing into a computerized control center.
>
> "Barnum and Bailey has got nothing on us," says Capt. Packard Mills, who oversees the TOC's operations.
>
> Outside, the tent and camouflage look typical old Army except for the

8. *Newsweek,* 11 March 1991, 17.

> satellite dish. Inside, video screens are coming to life amid the clutter. It's a long way from chalkboards and grease pencils. And a long way from battle-ready.
>
> "When you first set it up, it looks like a scene in Star Trek when you just got hit by the Klingons," Mills concedes.
>
> "We've got people tripping over cables and plugging into the wrong thing. It's keeping me busy," says a harried Sgt. 1st Class Tyler Vandesteeg. He's one of an emerging class: a Webmaster of war.

Cut and Paste: For the comfort of origins, I would say this investigation begins on a hilltop in the Mojave Desert, where I had been sent by *Wired* magazine to write about 'Operation Desert Hammer IV,' the first 'digitised' rotation of troops through the National Training Center at Fort Irwin. At the high end of the lethality spectrum there was the improved M1A2 Abrams main battle tank, carrying an IVIS (Inter-Vehicular Information System — 'Knowledge is Power' says the brochure) which could collect real-time battlefield data from overhead JSTAR aircraft (Joint Surveillance and Target Attack Radar System), Pioneer unmanned aerial vehicles equipped with video cameras, and global positioning satellite systems (GPS) to display icons of friendlies and foes on a computer-generated map overlay. At the low end, there was the '21st Century Land Warrior' (also called 'Warfighter,' but never 'soldier' or 'infantryman'), who came equipped with augmented day and night vision scopes mounted on his M-16, a GPS, 8-millimetre video camera and 1-inch ocular LED screen connected by a flexible arm to his kevlar, and a 486 Lightweight Computer Unit in his backpack, all wired for voice or digital-burst communication to a BattleSpace Command Vehicle with an All Source Analysis System which could collate the information and coordinate the attack through a customised Windows programme. 'Using the power of the computer microprocessor and digital electronics', digitisation was designed to be a 'force multiplier': the 'horizontal integration of information nodes' and the 'exchange of real-time information and data' was going 'to establish friendly force dominance of enemy forces.' In short, the Army was a creating a C^4I bundle (command, control, communication, computers and intelligence) of soft-, hard-, and wetware for the coming information war.

But up on that hilltop, as the simulated battle began at dawn with Black Hawks and Apaches flying so close to the deck they were below us, F-16s and A-10s roaring overhead, followed by the dust and smoke trails of M1 tanks, it was difficult to tell just what was going on. Our personable handler, Major Childress, former commander of an OPFOR unit and now head of public affairs at the NTC, did his best to explain, providing a running commentary for what we could see — and also what we could hear as we eavesdropped on the radio traffic among the combatants. Accounts of confusion and in more than one instance, fratricide or 'friendly fire', were overheard. But it was an aside from another member of the press that was to provide some much-needed historical perspective. For the most part my media cohort avoided me. I would like to think it was because of my intelligent questions and refusal to suck up to the brass, but it was more likely my failure to observe the press dress-code of Banana Republic safari vests, surplus fatigue pants, and desert jump boots. But at that moment, Austin Bay, ex-Army, military historian, and co-author of *A Quick and Dirty Guide to War*, turned to me and said: "It's just like Salisbury Plain." I knowingly nodded, but before I could ask what this had to do with lunch, we were interrupted by 'Krasnovians' in simulated T-80 tanks, who were about to overrun our perch as they outflanked the 24th Mechanized. We got the order to move, and during a dash through the desert in an humvee, Bay filled me in. Salisbury Plain was the British forerunner of the NTC, and it was there in the 1920s that troops, tanks, and airplanes, aided by wireless, came together for the first coordinated demonstration of mobile armored warfare. It was, said Bay, a revolution.

Flashforward. A few years later, I had the opportunity to check out his story. Killing time at the Bodleian Library at Oxford, I began searching the microfiche roles of the *Daily Telegraph*, not so much out of curiosity about the event as how it was reported: Was it recognised as a revolution at the time? I chose the *Telegraph* because I knew that Liddell Hart had been its military correspondent — and much more.[9] Hart, a decorated officer during the First World War, had made a name for himself as an early proponent for mechanisation, for a 'New Model' army based on 'tank marines' ready to use 'the indirect approach,' to fight highly mobile battles on land as the navy fought at sea. At a time when Germany was disarming under the agreements of the Treaty of Versailles, and the French, under the direction of war minister Andre Maginot, were re-casting trench warfare and protecting falling birth rates by a defensive frontier of concrete, the British had the luxury (no real enemy threat), the temperament (no desire to repeat the slaughter of the previous war), and the technology (still the leader in industrial innovation) to experiment.[10] From August 1927 to 1931, Salisbury Plain became the premier laboratory of a new form of warfare. Armored cars, light and medium tanks, motorized artillery, infantry in trucks and half-tracks, and even the odd horse were on the move, first during the day, later even at night. Hart's initial reports on the first exercises in 1927 were somewhat disdainful; aircraft were simulated, coloured flags stood in for anti-tank guns, and radios, where in evidence, rarely worked. But by the 'Armoured Force' exercise of 1928, the tone begins to

---

9. In a footnote — the *only* footnote — in his most popular book, *Strategy*, Hart's contribution to the Salisbury Plain exercises is acknowledged: "The strategy and tactics of the Mongols are dealt with more fully in the author's earlier book *Great Captains Unveiled* — which was chosen for the first experimental Mechanized Force in 1927." See B.H. Liddel Hart, *Strategy*, second revised edition (New York: Signet, 1974), 62.
10. See B.H. Liddell Hart, *Paris, or The Future of War* (New York: 1925); and Brian Bond and Martin Alexander, "Liddell Hart and De Gaulle: The Doctrines of Limited Liability and Mobile Defence," *Makers of Modern Strategy from Machiavelli to the Nuclear Age*, Gordon Craig and Felix Gilbert, eds., (Oxford: Clarendon Press, 1986), 598–623.

change. One hundred and fifty wireless sets were used for a maneuver which left an assembled group of brass and Members of Parliament highly impressed. Hart considered the exercises a success in 1931, when the 1st Brigade Royal Tank Regiment, taking orders by radio, managed to maneuver through the fog in concert to arrive on time before a gathering of the Army Council.

Reopen USA Today Online, "Cybersoldiers test weapons of high-tech war"

> Assuming the digital force passes muster, the Army could soon be asking Congress for lots more money. The General Accounting Office estimates it would cost $4 billion to outfit all 10 active-duty divisions. But Maj. Gen. Robert Scales, author of the official Army history of the gulf war, warns it will be money ill-spent if all the Army does is perfect war as we know it today. That's what France did after World War I with its Maginot line, a defence easily overrun because it failed to anticipate Adolf Hitler's high-speed mechanized attack.

From paper to film to screen to keyboard to screen to disk to screen to keyboard to disk to paper. What follows is a brief sampling of his fluent — and influential — accounts in the *Daily Telegraph* of the first exercises on Salisbury Plain in 1927. On the front pages were stories about the Naval Conference in Geneva (most notably, friction between the U.S. and Great Britain — with Japanese support — on cruiser tonnage and gun size), death sentences for Nicola Sacco and Bartolomeo Vanzetti, Italian anarchists, "Trotsky's Victory — Stalin's Move Checked — Surprise for Moscow," 'a world not ripe for disarmament.' Hart's early articles were on page five or later, mixed in with military bands and tanks bogged down in the mud; gradually the articles moved up to page one. Entertainment is liberally mixed with education. They read like the bread and circus of late empires — much like our own evening news.

The Daily Telegraph, Monday 1 August 1927, "Tidworth Tattoo — Modern War Staged,"

> Tidworth is the home of the mechanized force which is expected to play a great part in the future development of the Army. Therefore it is fitting that the star attraction of the Southern Command Tattoo, which commenced before many thousands of people in the arena in Tidworth on Saturday night, should be a 'battle' in which the latest mechanized units take part. When an interesting programme was nearing its end, the searchlights flashed on to an Eastern fort, where picturesque Eastern marauders were taking rest. Almost immediately the battle began. A signal for assistance sent by the British commander brought a reconnaissance car to the spot, and, following quickly in its wake, came the mechanized machine guns, the latest swift-moving tankettes spitting fire, with a self-propelled gun giving protection to the British force, and in doing so adding to the din. The mobility of the new armoured units enhanced the realism of the episode, and undoubtedly this battle will prove one of the most attractive features of the performances.

> There is plenty of variety in the programme, for following community singing and the fanfare of trumpets, massed bands of the 2nd Cavalry and 7th Infantry Brigades enter the arena in peace-time uniform, the cavalry bandsmen mounted, and all playing delightful music. ... Lancer trick riders carry through amazing feats and some remarkable jumping, the obstacles including a donkey and cart, bed, fire hoop, and fire bar. ... The concluding item before the reassembling of the soldier actors is a display by the Royal Air Force in illuminated aeroplanes ...
>
> The tattoo was a huge success on its first night and will be continued during the week ... the railway companies are running excursions from all over the South of England and buses are expected to bring many hundreds of spectators.[11]

Flash sidewise. To "Hearing a Face — Television Broadcast," *Daily Telegraph* article next to Hart's first account of the 'Tidworth Tattoo', 1 August 1927.

> Giving a broadcast lecture at the British Empire Exhibition at Edinburgh on Saturday night, Mr. J.L. Baird, the inventor of television, said he had asked three chance acquaintances the meaning of the word 'television.' One said that it was an island off the Coast of Africa, the second that it was a form of telepathy, and the third that it was a kidney disease. Television meant actually seeing by wireless. The scene was first turned into a sound, which was then broadcast, and turned back into an image at the receiver. Every face had its own particular sound.
>
> A phonograph record was then played on which the television sound of Mr. Baird's face had been recorded. It sounded something like the rasp of a file with a peculiar rhythmic whistle underlying it. This was broadcast by the B.B.C., so that listeners for the first time in history had the opportunity of hearing what a face sounded like. The lecturer went on to describe his discovery of television, and said that the first person ever seen by television was an office boy, who had to be bribed with 2s 6d to submit to the experiment. The latest development of television had rendered it possible to see in total darkness, invisible rays being used. Steady progress was being made in developing the invention to a commercial stage, and he hoped that television would very shortly be available to the general public.[12]

Historical note. One year later, the same year that motorised and wireless transmissions were linked in simulated warfare, similar breakthroughs in television were made by engineers at General Electric. From experimental station 'W2AXAD' they broadcast the second-ever television image, about the size of an index card. What did they choose to broadcast? A simulation of a missile attack on New York City. The point of view was from the missile, a flight ending in an explosion, then nothing.[13]

---

11. "Tidworth Tattoo — Modern War Staged," *The Daily Telegraph*, 1 August 1927, 6.
12. "Hearing a Face — Television Broadcast," *The Daily Telegraph*, 1 August 1927, 6.
13. Eric Barnouw, *A Tower in Babel: A History of Broadcasting in the United States,* vol. I

Hammering the point of the missile home. Paul Virilio, in his preface to the English edition of *War and Cinema*, writes: "A war of pictures and sounds is replacing the war of objects (projectiles and missiles). In a technicians' version of an all-seeing Divinity, ever ruling out accident and surprise, the drive is on for a general system of illumination that will allow everything to be seen and known, at every moment and in every place."[14]

Daily Telegraph, "Tanks 'In Action' on Salisbury Plain — Bombing Attack Thrills — Triumph for Mechanised Army," by Captain B.H. Liddell Hart, Tidworth, Friday Night, 20 August 1927.

> To-day the training proper of Mechanised Force was inaugurated on Salisbury Plain. More definite localization is impossible because the exercise covered too wide an area, and that is a point to the good, for an immense broadening not only of space but of mental horizons is the only way in which full value can be obtained from these tentative experiments in mechanized warfare ...
>
> Further, one discovers that common-sense again has overcome another of the apparent drawbacks of the Mechanical Force as originally constituted — by distributing this mechanical potpourri into groups according to the qualities of the mechanized units which compose it. Thus for marches it is divided into a fast group, comprising only the Armoured Car companies, whose normal rate of march is reckoned at 25 m.p.h.; a medium group, comprising light batteries, field companies R.E., machine-gun battalions, and mechanized transport, all conveyed in semi-track or six-wheeled vehicles, whose normal rate of march is reckoned at 10 m.p.h.; a slow group, composed of tanks, tankettes, and mechanised artillery, whose normal rate is reckoned at 7 m.p.h.. These normal rates are, of course, on the conservative side, with the sound object of allowing for hindrances and reducing wear and tear.
>
> 'Attacks from the Air'
>
> To-day's test was a 'peace march by day,' a phrase which implies not that it was under peace conditions, but that it was conceived as taking place in rear of the advanced forces and so not likely to encounter the enemy. But the serpentine column which wound its length in coils over a distance of some thirty-two miles, suffered bombing and machine-gun attacks from the air, when checked by road blocks, and had also to pass through an imaginary gassed area — presumably by enemy aircraft.
>
> 'The First Test'
>
> ... [T]he umpire created an impassable block by the declaration that a vehicle had broken down — or been blown up. It was about as severe a test as could have been imposed on a new-born force, and it was not surprising that twelve

(New York: Oxford University Press, 1966), 231.

14. Paul Virilio, *War and Cinema: The Logistics of Perception,* trans. Patrick Camiller (New York: Verso, 1989), 4.

> minutes elapsed before the first vehicle moved off the road to lead the way along an alternative route hurriedly reconnoitred on the flank of the sunken road a route which, incidentally, enabled the whole column to display their cross-country ability and to get on to another and wider road altogether. ... At this juncture flights of single-seater Gamecocks came diving over the trees and swooped down on the vehicles in the road, dropping 'flour' bombs and firing their machine guns. So low did they come, so spectacular was their handling of these bullet-like machines that the spectators had a series of thrills. That they would have caused heavy casualties there is little doubt, but even in a stage-managed affair they were late enough behind time to lose several minutes of their opportunity, and in war, given greater practice by the mechanised troops, it is not often that such an ideal obstacle, such a heaven-sent breakdown, and the appearance of the enemy aircraft would all three combine ...[15]

'Mechanical Gods' of Modern Warfare — Tanks in Night Move — Driving feat in the Dark, Hart, 23 August 1927.

> Between 10 p.m. last night and daybreak this morning, the Mechanised Force, under Colonel R.J. Collins, carried out the second of its trial schemes — a night march along some fourteen miles of rain-steeped roads to a rendezvous near the Bustard, north-east of Shrewton, and a return in two columns across country. ... If the test was severe for a newly-assembled and still inchoate force, the condition increases its severity to such a pitch that even an ardent believer in mechanisation was astonished at the practically hundred percent fulfilment which was achieved.
>
> 'Primeval Monsters'
>
> I watched the column from a point close to Stonehenge, and in the apt and eerie setting of that dreary monolith-surmounted down, at midnight, little imagination was needed to picture it as the passage of a herd of primeval monsters or legendary dragons, with glassy eyes shining in the darkness, fiery breath, and scale-coated body. So irresistible was the impression that I pity any belated motorist who met them, unprepared on his homeward road. And the passage by Stonehenge had also a symbolical effect, for there the gods of the prehistoric past could be conceived as watching from their long-abandoned altars the procession of the mechanical gods of modem man — both equally the creation of man, but the one expressing the static mentality of the past, and the other the ever-changing, restless motion of the mind of to-day.[16]

Historical sidebar: The next day, Sacco and Vanzetti were electrocuted.

Impressed, but not convinced, the British general staff failed to learn the lessons of armored warfare wargamed on Salisbury Plain. Defeated, and some might even say rendered desperate by disarmament and the fiscal restraints

15. "Tanks 'In Action' on Salisbury Plain," *The Daily Telegraph*, 20 August 1927.
16. "'Mechanical Gods' of Modern Warfare," *The Daily Telegraph*, 23 August 1927, 11.

imposed by reparations, the German staff did not. They carefully studied Hart's writings as well as Brigadier Charles Broad's 1929 booklet, *Mechanized and Armoured Formations*, which conceived of the tank not as a support for infantry but as a fast-moving independent force that could create shock, chaos, and demoralisation in enemy forces. In 1939, they applied those lessons with spectacular results.

Exterior. Bodleian. Film-noirish voice-over: Why, when Austin Bay looked down from the desert hillside, did he see Salisbury Plain from his desert perch, rather than Poland or France, collapsing under the speed and fury of the Panzer *Blitzkrieg*? And if we were the British, who then, or rather, who now, are the Germans?

Travelling shot. It would take a few more manoeuvres in the field before I would find an answer. Two tours at Fort Irwin; two trips to the annual Interservice/Industry Training Systems Conference in Orlando, where simulation industries like Boeing, Lockheed, Loral, Silicon Graphics, Evans and Sutherland paraded their wares to their military; Central Command in Tampa, on the heels of Schwarzkopf and the wargame that took us to Iraq, Internal Look 90; Fort Knox, home to our dwindling gold supply and to the ultramodern tank SIMNET; Hohenfels, Germany, to observe the lst Armored Division as they 'peacegamed' their intervention into Bosnia; Advanced Research Projects Agency (ARPA) in Virginia, to learn how the Synthetic Theater of War (STOW) was created to integrate virtual, live, and constructive simulations of war in real time; and, finally, the Office of Net Assessment at The Pentagon, where I found the Yoda of the 'Revolution in Military Affairs.'

Outtakes. In many ways, the itinerary for the journey was determined by an air-express package that I received from the Office of the Secretary of the Army the day before I was to leave for Fort Irwin. Although it did not come with an acronym de-coder ring, I was able to make some sense of it. Officially, it was identified as the press kit for the Advanced Warfighting Experiment ('AWE'). But this did not do it justice. Collected in a large three-ring binder with the triangle logo for 'The Digital Battlefield' on the cover — satellite, helicopter, and tank in each corner, connected by lightning bolts to a Warfighter in the middle — were over thirty press releases, brochures, and articles on the Army of the future. In style and content they replicated the corporate publications that I had picked up three years earlier in Orlando. Computer-generated images were mixed in with all kinds of fonts and graphics. Indeed, it all looked a bit like *Wired*.

Leading the paper charge of the simulation brigade was a prolegomenon from the office of the Chief of Staff. It bears quotation, not just for its Toffleresque rhetoric, but for its encapsulation of the rationale behind the 21st Century Army, Force XXI:

> Today, we are on a threshold of a new era, and we must proceed into it decisively. Today the Industrial Age is being superseded by the Information

> Age, the Third Wave, hard on the heels of the agrarian industrial eras. Our present Army is well-configured to fight and win in the late Industrial Age, and we can handle Agrarian-Age foes as well. We have begun to move into Third Wave Warfare, to evolve a new force for a new century — Force XXI.

A series of categorical imperatives for the Force XXI follow. They call for nothing short of a paradigm-shift:

> Force XXI will represent a new way of thinking for a new wave of warfare. We must be strategically flexible and more lethal. We must leverage the power of the best soldiers in our history through the use of state-of-the-art simulations and realistic, simulator enhanced training. We must accommodate the wide-range of operations being demanded of us. Intellectual change leads to physical change — the mental shift goes before the software and hardware.

One brochure, slicker than all the rest, maps out how the Army was making the future present. It bears the short, yet pretentious, title: "The Vision." It leads with the now common litany of the national security mandarins, that with the fall of the Berlin wall, the dissolution of the Soviet Union, the rise of regional powers, and the advent of MTV (reading between the lines here) no one can safely predict what is to come, nor who is to be the next enemy. The Chief of Staff, General Gordon Sullivan, asks, "What's next?" and answers, "No one knows." Therefore, "We are relatively safe in predicting, however, that the strategic environment in the next decade will be dynamic, uncertain, and unstable." Military jargon married to 'technospeak' usually calls for high waders, so I was surprised to find, a few pages later, a box in the section on 'Exploit Modeling and Simulation' that read, well, like a good cyberpunk novel:

> The Distributed Simulation Internet, projected for the turn of the century is a creature of another order entirely from SIMNET. Ten thousand linked simulators! Entire literal armies online, Global real-time, broadband, fiber-optic, satellite-assisted, military simulation networking. And not just connected, not just simulated. Seamless.

It gets better, and for good reason: it was written by Bruce Sterling for *Wired*. What does it mean when *Wired* is appropriated for the Army's 'Vision'? Perhaps in the idea-void of post-Cold War strategy, shortly after 'enlargement' (of democracy and free markets) is offered by the Clinton Administration as a plausible foreign policy replacement for 'containment' (of the 'Soviet Threat'), it is wholly understandable that the Army's visionary reach should exceed its rhetorical grasp. Indeed, I had come across much stranger intertexts in the course of the visit to Fort Irwin. One briefer had described the intensity of Desert Hammer as somewhere between the Gulf War and *Red Storm Rising*. Not such a surprise, considering that former Vice President Quayle had once defended Star Wars (the anti-missile system, not the movie) by citing the same Clancy novel.

Or perhaps something else was going on; something I sensed at the NTC when I was in the M1A2 tank, and again when I was granted videotaping access not once but twice to the Star Wars building, command central of the NTC from which the battles are run and to which the signals from hilltop remote video cameras and overhead reconnaissance are beamed for the after-action videotape reviews. Was my presence at Fort Irwin — no less so than Bruce Sterling's in 'The Vision' — just one more tactical exercise in the Army's much-vaunted Information War? Was journalistic simulation one more front for the successor to Salisbury Plain?

As early as 1964, after reading a breathless promotional account of the 'Cyborg' under development by General Electric and the military (from the photographs it looked like a robotic elephant), Lewis Mumford warned of the coming of a new 'technological exhibitionism.' Thirty years on, was I bearing witness to an even more powerful, possibly perverse hybrid? What happens when you combine media voyeurism, technological exhibitionism, and strategic simulations? News flash: In the 21st Century Army, you get the cyber-deterrent.

If this sounds far-fetched, remember the worst-case scenario that currently underlies strategic thinking. As CIA director James Woolsey put it at his confirmation hearings, a "bewildering variety of poisonous snakes" has sprung forth from the slain dragon. With the dragon went the mighty, if mainly illusory, deterrence value of nuclear weapons. On a quest since Vietnam (to fight only quick, popular, winnable wars), and imbued by the spirit of Sun Tzu ("Those skilled in war subdue the enemy's army without battle"), the 21st Century Army has perhaps now found in the cyber-deterrent its Holy Grail. It is fast, digitised, and as spectacular in simulation as it is global in surveillance. The digitised option also has the advantage of being out of reach of all but the richest 'rogues.' And it makes a hell of a 'photo-op.'

Moreover, the digitized deterrence machine bears an important similarity to its nuclear counterpart: it does not *necessarily* have to work in order to be effective. Its power lies in a symbolic exchange of metaphysical signs — give or take the odd reality-check in the desert to bring religion to the doubters. Hence spectacles like Desert Hammer IV, to render visible and plausible the cyber-deterrent for all those potential snakes that might not have sufficiently learned the over-hyped lessons of the first (if prototypical) cyberwar, Desert Storm.

Once again the desert functions as backdrop for the melodrama of national security. With an assist from Disneyland, Hollywood, and Silicon Valley, the National Training Center, full of video cameras, computerized special effects, not to mention thrilling rides, has superseded Los Alamos and the Nevada Test Site to become the premier production set for the next generation of U.S. strategic superiority. Can the Army go on to win the information war without firing another (real) shot? Of slightly lesser concern, can one conduct a critical inquiry of the information war without becoming, well, just another informant for it, a box in the Army's sequel issue, '(Re)Visions'?

Cut and run (and paste). Combat and Maneuver Training Center (CMTC), Hohenfels, Germany.

The U.S. Army owns, or more precisely, has 'manoeuvre rights' over a significant piece of real estate in southern Germany, 178 square kilometers worth in Hohenfels alone. Spread out over the State of Bavaria like an isosceles triangle are the three major sites of the U.S. Seventh Army Training Command, through which the European-based U.S. troops, as well as some units from the British, Spanish, Canadian, and German armies and the Dutch marines, rotate through for some laser-simulated warfare as well as for live-fire exercises. The centres have an interesting heritage. Grafenwoehr, the oldest, was set up by the Royal Bavarian Army in 1907 to 'play' some of the earliest *Kriegsspiele*, or war games. It served as the southern tactical arm of the northern Prussian head, most infamously represented by Count von Schleiffen, Chief of the General Staff, who, in 1905, designed the famous Schleiffen Plan that was supposed to anticipate the next German conflict. Instead, its iron-clad 'war by timetable' helped to precipitate the First World War as one mobilisation triggered a cascade of others throughout Europe. The two other training centres owe their origins to Hitler's rejection of the Treaty of Versailles, the peace of the victors of the First World War which included the humiliating 100,000 troop limitation for Germany. Rapidly filling up the ranks with new conscripts, the *Wehrmacht* found itself short on training space. Grafenwoehr was expanded, and two new sites were created: Wildflecken in 1937 for the IX German Corps, and Hohenfels in 1938 for the VII German Corps. It was here that the lessons of Salisbury Plain were applied.

The morning I drove past the front gate and into the Hohenfels Combat Manoeuver Training Center, I learned a less-known part of its history. The tank-crossing sign, resembling more First World War lead toys than the Ml behemoths that skidded up the hill ahead of me, momentarily caught my attention. But it was the more conventional warning sign for "Cobblestones: Slippery When Wet" that seemed out of place. I later asked my handler, the very smart, very affable Colonel Wallace, why the short strip of quaint cobblestone interrupted the finely graded, modern asphalt road into the base. He thought it had been left intact as a tribute to the Polish construction workers. Later I filled in the blanks: Hohenfels, begun in 1938 and finished in 1940, had evidently been built by Polish *sklavenarbeiter*, slave labour. Wars, when gamed, tend to lose their history of blood and deception: "Slippery When Wet" joined 'Trust no one' as my coda during my visit to Hohenfels.

The reason I was there had taken on a special urgency. Two weeks before my arrival at Hohenfels, NATO air strikes on Bosnian Serb ammunition dumps triggered the hostage taking of over 300 UN peacekeepers. The 'cold peace' flared hot when French soldiers in Sarajevo fought back after Bosnian

Serbs disguised in French uniforms and UN blue helmets tried to take the Vrbanja Bridge. Britain and France announced plans to send a rapid reaction force: debate ensued as to whether it would be under UN command — and whether the new artillery, armoured vehicles, and helicopters would be painted UN-white or sovereign-camouflage. President Clinton, breaking with the stated policy of only providing U.S. troops in the event of best and worst case scenarios — to monitor a peace accord or to cover a UN withdrawal — suddenly announced that he was ready to 'temporarily' send troops in support of the British and French forces. But morning-after polls and the shootdown of a U.S. F-16 pilot by the Serbs quickly reversed that readiness. In fact, as I drove through Hohenfels for my morning briefing I spied in the *Stars and Stripes* newspaper box in front of the PX Burger King a tall headline and a big photo: "A Hero's Welcome" ... "Air Force Pilot Capt. Scott F. O'Grady looks mighty glad to be back — alive — at Aviano AB."

It seemed like the right time to come to Hohenfels to observe an 'Operation other than War.' Just what that meant was supposed to be the subject of the morning brief. But there was some initial confusion, not least because somewhere between the time of my fax-barrage requesting a visit to the base and my arrival, a name-change had taken place. 'Operations other than War' had been replaced by the more anodyne 'Stability Operations.' Word had not quite made its way through the ranks, and people kept shifting back and forth between the two. The confusion mounted as I sat in a darkened theater with my two handlers, Captain Fisher and Colonel Wallace, and listened to the opening to Major Demike's multimedia, name-negating brief. The Major was clearly in a 'take-no-prisoners' attitude toward the English language: 'Army units from USAREUR (troops in Europe) rotate through the CMTC (I got that one) at least once a year for 21 days of Force-on-Opfor training' (good guys vs. bad guys), 'situational training with MILES in the Box' (dial-a-scenario field exercises using lasers rather than bullets), 'BBS training' (not bulletin-board systems, but networked computer battle simulations with units based elsewhere), and 'after-action reviews' (video presentations of what went wrong on the battlefield).

It was all very impressive, but after five years of research on wargames and probably one too many jarring rides in a humvee, I had just about reached my tolerance for military speak. I had gone one brief too far, and I was ready to go in search of that faceless desk-jockey sitting somewhere in an inner-ring, windowless office of the Pentagon, whose sole mission was to regularly abbreviate and, if necessary, change the name of anything in the military that becomes decipherable to the layman before its half-life of usefulness is over.

But it would have to wait until after the 'mother of all techno-briefs.' Major Demike got into it with vigour: "We have at CMTC the most realistic battlefield. The instrumentation system is state of the art. No other training centre in the world has an instrumentation system like ours. It is the best in the world." He skipped through technology like the MILES (Multiple Integrated Laser

Engagement System) for firing and recording laser hits, the microwave relays which allowed for near real-time production of the video after-action reviews, and the simulated mortar and artillery fire. To punctuate the point, Colonel Wallace stepped in: "Once a unit goes into the Box, with the exception that they're shooting laser bullets, and that a guy, instead of falling down with a gunshot wound, will read from a card he's carrying in his pocket how badly hurt he is, virtually everything we do is real. There's nothing simulated in the Box."

The Major became more animated when he moved into the details of the technological capability of the CMTC. Instrumentation systems gather and process battlefield data that observer/controllers use to provide instant feedback for both sides of the operation. There is a seamless web of command and control between Building 100 (the 'Star Wars') centre from which the battles are run, and the troops in the Box. For instance, simulated artillery attacks are launched via Sun Microsystem work stations, and hits are assessed according to probability software which calculates trajectories, terrain, and the grid locations of vehicles and troops which are constantly updated by Global Positioning Systems. Hits are then transmitted to each vehicle, as a 'commo kill' (communications knocked out), near miss, or 'catastrophically destroyed.' News of a simulated death comes in a female voice: the female voice gets the attention of the adrenalised or battle-fatigued soldier. My query about what happens when women eventually get to join in on the combat simulations was met with a blank stare by the Major, but the Colonel picked up on it: the female voice will always stand out from the background of male ones. My stock question about the realism of the simulated battlefield received the stock answer, but with a raising of the technological ante: the National Training Center, CMTC's better-known state-side rival in the Mojave desert (see *Wired*, 2:09) was still using the first generation of MILES to simulate weapons effects, while they had the interactive MILES 2 with data communication interface ($9,000 a unit). 'Everything is wired' said the Major, who clearly had an ear for a soundbite.

After a long slog through computer graphics on the organisation and function of the CMTC, we finally got to the geopolitical gist of tomorrow's 'Stability Operation.' Up came a map of 'Danubia,' trisected into 'Sowenia,' 'Vilslakia,' 'Juraland,' and, looking like very much like a small fiefdom among them, the CMTC. The Major's pointer started to fly: "Three separate countries have split off from Danubia — Sowenia and Vilslakia are at odds with each other. When we want to transition into high-intensity conflict, we have Juraland, which has heavy forces, come in on the side of one or other of the parties." Prodded to just once utter the word 'Bosnia,' he would go no further, except to say that the scenario was based on intelligence sources, CNN reports, and the 'threat books.' But for my benefit he did add, "You don't have to be a rocket scientist to figure out what this is modelled on."

No rocket scientist, I resorted to a kind of semiotics to sort out the countries. The new countries of the disintegrating Danubia bore some obvious similarities to the region of Yugoslavia: to the former republic, now independent state of Slovenia, or perhaps the western enclave of Slavonia contested by the Croats and Serbs; and, of course, to the Jural mountain range. 'Vilslakia' remained a mystery. The countries surrounding Danubia were familiar enough that I sought out my own intelligence source, Microsoft's CD-ROM version of *Cinemania '95*. It was not needed for the country to the northwest: 'Teutonia' referred back to the early Germanic tribes. However, 'Freedonia' to the northeast of Danubia was clearly taken from the 1933 war satire *Duck Soup*, in which Groucho Marx so effectively played the power-hungry dictator of said-country that the real dictator Mussolini banned the film from Italy. And below Danubia was 'Ruritania,' the country in the clouds which provided the surreal setting for W.C. Field's 1941 classic, *Never Give a Sucker an Even Break*. What should one make of the Army's strange choice of simulated countries? Probably nothing much, except that some wargamer had a sense of humour as well as of history. But I was left wondering: play by the intertext, die by the intertext?

The briefing ended with a short video of a 'Stability Operation.' By way of introduction, Colonel Wallace informed me that "none of this stuff is staged, it's all from live footage taken by the Viper video teams in the Box." Before I can fully enjoy the Colonel's knack for paradox, the lights dim, the screen flickers, and Graham Nash is singing something about 'soldiers of peace just playing the game.' The first clip is of a confrontation between partisans and soldiers that escalates into heated words; the last is in the same tent, with handshakes and professions of friendship being exchanged. In between, UN convoys are stopped by civilians, soldiers go down, wounded or dead, a body-bagged corpse is spat upon by a partisan, food supplies are hijacked by townspeople, a female member of the media gets shoved around, an explosion and panic in the town streets, a sniper fires on a humvee, dogs sniff for explosives, infiltrators are caught in a nightscope, a UN flag waves defiantly, and an old man drops to his knees in the mud in front of a humvee, begging for food. More in the sentimental aesthetic of an AT&T 'advert' than an hyperreal MTV clip, it is strangely moving. I am disarmed by it.

But the mood shifts quickly when the Major concludes the briefing by handing me a three-inch thick pile of documents. The rest of the day was a whirlwind of briefs-to-go. First stop was the 'Warlord Simulation Center,' full of desktops and more Sun Microsystems for planning, preparing, and running simulations in the Box, out of the Box, or through the cyber-Box — that is, simulation networking (SIMNET), 'remoting via satellite in and out of the Box to anywhere in the world.' Next stop was a cavernous warehouse, full of MILES gear under the watchful eye of Sergeant Kraus, who probably gave the best brief of the day. A man who clearly loves his job — or just eager for some human company — he was as articulate as his lasers ("instead of a bullet it sends out 120 words on a laser beam, in the center are eight kill words, anything else is a wound or near miss"), as he made his way through the various shapes, types, and generations of laser and sensors, all set up on a variety of weapons and menacing mannequins. He was stumped only once, when I asked what would happen if a Danubian snuck up and hit one of his dummies on the head. Would any bells and lights go off? "Excuse me?" he said. "ROE?" Colonel Wallace intervened to explain: "Against the Rules of Engagement. One meter rule. No physical contact in the Box." It seems that one conveys body-to-body harm with real words, not laser words — i.e., 'I am butt-stroking you now, so fall down.' I would later find out that in 'Operations other than War,' the Rules of Engagement were there to be broken.

The day ended with an interview with the pugnacious commander of the base, Colonel Lenz, who made a persuasive case for Stability Operations as essential training for the increasing number of missions in that 'grey area between war and peace.' He would not, however, be drawn out on the relationship between Stability Operations and Bosnia, especially when I queried him about the possibility that some might find the notion of stability based on the status quo to be offensive, in both senses of word, when stabilisation is perceived to be an enemy of justice, or simply just desserts. "That's above my pay-grade." At the end of the interview he kindly suggested a de-brief after my visit to the Box: "I've got people upstairs who can suck a guy's brain dry."

That was sufficient incentive to stay up that night and wade through the stack of papers that I had been given. The bulk of it was a four-hundred page document called the "Coordinating Draft of the 7th Army Training Command White Paper of Mission Training Plan for Military Operations Other Than War." A substantial part of it breaks down the 'Critical Tasks of the Task Force,' like the establishment of a quick reaction force, checkpoints, lodgments; conduct liaison with local authorities and convoy escort operations; provide command and control and protect the force; and, of no lesser importance, plan for media. Specific scenarios for battalions, companies, and platoons are spelled out. The philosophy of Operations Other Than War is conveyed in the introduction, and after wading through all the acronymic muck and bureaucratese ("Traditional MTP crosswalk matrixes for references and collective tasks are also included in this MTP") — the final paragraph emerges as a reasonably clear summary of the purpose of the plan:

> As we continue to maintain our proficiency in traditional wartime operations, our forces must also be ready to operate effectively in non-traditional roles. Units involved in conflicts anywhere within the full spectrum of operations will always face some elements of a complex battlefield. These elements include civilians in the area of operations, the press, local authorities, and private organizations. This White Paper is designed to assist leaders at all levels to more fully understand and prepare for these new challenges.

In other words, this 'White Paper' was this year's model for the hi-tech, post-Cold War simulations and training exercises which would prepare U.S. armed forces for pre-peacekeeping non-interventions into those post-imperial spaces where once- and 'wannabe'-states were engaged in post-war warring. In terms of past experiences rather than future threats, Somalia, Haiti, Rwanda, and — judging from the many references to the British Wider Peacekeeping Manual — Northern Ireland lurked between the lines. But in this simulated shadowland between military combat, police action, and relief aid, other ghosts could be discerned: Bosnia, yes, but why not as the next operation other than war, a counter-narcotics operation in Mexico? Or a quarantine of a paramilitary survivalist camp in Idaho? Or checkpoints and convoy escort through a persistently riotous Los Angeles? This week, however, the enemy at Hohenfels reflected the headlines.

Very early the next day I was heading for the Box, where the warring ethnic groups of a disintegrating 'Danubia' were about to make life very hard for the visiting 1st Armored Division. The next morning began with a low fog — confirmed by the weather report provided at the 'Battle Update for Rotation 95–10.' The mission: 'to provide humanitarian assistance and separate belligerent factions.' Computer graphics were projected in meticulous detail, breaking the mission from the highest level of 'UNDANFOR' (United Nations Danubian Force) commander down to equipment lists, tactical rules of engagement, task force organisation, and maps with vehicle and troop positions. A schedule of major events was put up, some of which required translation, like '1100 — Scud Ambush of Convoy' (not the missile, but the 'Sowenian Communist Urban Defenders'), or '2230 — Jerk Raid vs. Care Facility in Raversdorf (again, not Steve Martin, but the 'Jurische Ethnic Rights Korps,' guerrilla forces operating in the south sector). I was in bad need of a scorecard.

Finally, we were on our way to the Box. There was a bit of delay as I struggled with the camouflaged ensemble of gortex jacket, pants, and boots (for the mud), and as my faith in our humvee was tested when the door handle came off in my hand. But Colonel Wallace proved to be as good a handle-fixer as he was a handler, and we were soon off. During the short ride through a gently sloping open terrain with trees on most of the hilltops, Colonel Wallace did the eco-army routine — "there are more trees and grass growing now than when we got here" — and as if on cue, a substantial herd of deer dashed across the road in front of us. The valleys and hillsides looked pretty chewed up by all the maneouvres, 'portaloos' dotted the landscape, but the fauna seemed to appreciate the fact that the U.S. Army — unlike the Bavarian hunters outside the Box — were shooting blanks.

The first stop was a UN checkpoint, one of many where civilians were stopped and forced to do a kind of 'self-search' for weapons or explosives. Most of the M1 tanks and Bradleys had their turrets reversed, the universal symbol of nonaggression (or surrender). We arrived with a UN food convoy which was supposed to pass through the mock-town of 'Ubingsdorf.' The town came complete with the steep-roofed houses of Bavaria, a church with a steeple (no sniper in sight), a cemetery (no names on the gravestones), a mix of 'Vilslakian' and 'Sowenian' townspeople (dressed by a retired psy-ops sergeant in what he described as 'the eastern-european, grunge-look,' accessorised with the requisite MILES vest), and a mayor in a green-felt fedora, who was insisting that the food be off-loaded for his hungry people.

Language differences, a belligerent crowd, an aggressive reporter with an intrusive cameraman, all jacked up the tension level. 'Lt. Colonel Vladimir,' commander of the local Vilslakian garrison, was refusing to bring the rabble to order. Chants for food in a kind of pidgin German — 'Essen, Essen' — made voice communication difficult. Suddenly, the crowd began to move toward the trucks, and a few rocks were thrown. The U.S. troops began to retreat back to the trucks, but already some of the townspeople were clambering up onto them. It was then that the first rule of engagement, right up there with the 'Prime Directive' of no-no's, was broken by one of the soldiers when he grabbed a civilian to toss him off. 'One meter rule, one meter rule!' was shouted by the observer/controllers on the scene. Some tanks and Bradleys, probably called up by the besieged sergeant in charge of negotiating with the mayor, came roaring up to join the convoy. The situation died down when the townspeople were rounded up and put under guard. Negotiations resumed, resulting in something of a compromise: the food would be unloaded at the local UN headquarters. But after the troops pulled out, I watched as some of the townspeople pulled off the most realistic maneouvre of the day: they scampered off with some of the large crates of food, evidently for their own purposes. Colonel Wallace later told me this was not in the script. I had witnessed some 'Box Improv.'

The script-writers clearly had it in for this convoy. At just about every checkpoint, food had to be traded for safe passage. And now, as we roared ahead in the Colonel's humvee for high ground, I noticed an observer/controller crouched in the ruins of a building probably dating back to the *Wehrmacht* days. A bad sign. As the convoy descended down the hill all hell broke loose — machine-gun fire from the hills, smoke bombs marking hits, and the light-and-sound show of MILES sensors going off. The M1 tanks and Bradleys reacted sluggishly to the ambush, not moving, and worse, keeping their turrets reversed in the defensive posture which made it impossible to identify the enemy with infra-red or thermal sights. Instead, someone called in for a Cobra helicopter gunship, breaking another rule of engagement: only 'minimum' or proportional force should be used in a counter-attack, to prevent a needless escalation of violence. From the last two engagements, it seemed apparent that the shift from war/sim to peace/sim was not going to be an easy one.

Salisbury Plains Forever. During my interview with Andrew Marshall, Director of the Office of Net Assessment, the intertext went into 'monalisa' overdrive. For five years I had managed to avoid the Pentagon. It was not just the reports of the rats in the basement that kept me away. I had learned the hard way that when it came to the Revolution in Military Affairs, the hype-to-reality ratio skewed as one went up the ranks. But at every pit-stop I got the same name, regardless of stripe on the sleeve or political colourings: go talk to Andy Marshall.

Several faxes later, I was sitting across from him in his spacious, paper-filled, very unmilitary Pentagon office. At 74 years old, Andrew W. Marshall has been around. Brought in by President Nixon, he helped set up the innocuous sounding 'Office of Net Assessment,' "to weigh the military balance in specific areas, what the important long-term trends are, and to highlight existing or emergent problem areas, or important opportunities that deserve top level management's attention to improve the future U.S. position in the continuing military-economic-political competition." His memoranda are legendary, and, for the most part, classified. They have ranged from broad politico-strategic issues like the decline of the Soviet Union, to no less important tactical debates about the advantages of sending Stingers to Afghanistan. The one that I kept hearing about bore a simple title — 'Some thoughts on Military Revolutions' — and was only seven pages long. When it was circulated on 23 August 1993, it was an idea in the wind; a year later, there were five task forces at the Pentagon alone, exploring the ramifications of the 'RMA' — the Revolution in Military Affairs.' He agreed to talk to me about it, on the record.[17] Here are some excerpts:

Der Derian: Could you tell me who you are and what do you do?

Marshall: Well I'll start with the history. I went to Rand at the beginning of '49 and I was there until the beginning of '72. Then I went and worked for Henry Kissinger at the National Security Council, and a couple of years later came here to set up this office.

I've been here ever since. This is the Office of Net Assessment and fundamentally what it tries to do is assess military situations with the intent of surfacing for the very top managers issues that they should pay attention to. I mean, based on emerging problems or opportunities. Of course, when I was first here we focused very much on the Soviet Union, and the more intense military, political, economic competition.

Now we really are working fundamentally on two things. One is exploring this idea of, you know, that the next twenty to thirty years may be another one of these periods where warfare changes in some very significant ways. We've done some earlier analyses of that before but, began about four years ago a much more intensive effort. I suppose we really began in '89 or so, and put out a preliminary assessment in July of '92 and have

17. Interview, Andrew Marshall, 21 June 1996.

been pursuing a variety of activities to try to understand the potential character of the change, to better understand the actual nature of what the change might be.

So that is one thing we are doing and the other is trying to take a very long term view of Asia and where it may go, again, over the next twenty or thirty years.

Der Derian: Would you call it a revolution or not?

Marshall: Well, I mean, we have picked up this terminology of revolution and, I think, at one level, or in one way, that's appropriate. It turns out that tactically it gets you into a lot of arguments you don't really need to be in about whether it is a revolution, or what things can be called a revolutions. Anything that can happen over a couple of decades can't be called a revolution, for some people ...

Der Derian: Would you call it a revolution?

Marshall: Yeah, I would ...

Der Derian: Why?

Marshall: Well I think, again, if you look back, there is all this historical work that people have done on, way back to the fifteenth century, looking at periods where over the course of, you know, couple of decades or so many new forms of warfare emerge that just dominate whatever was dominant before and that seems, you know, reasonable to call it a revolution. It was the Russians that first brought it to my attention, the writings that they began to put out in the late seventies and early eighties.

Der Derian: You mean your counterparts in Russia were speaking about a military revolution?

Marshall: Well, yes, beginning in the '70s and on into the early '80s they began talking about the fact that we were entering, or that the world was entering, another period of what they initially called a 'military technical revolution.' And they cited two previous periods as exemplars. One was the twenties and thirties where you get the big change in many areas of warfare, because of, well, in some ways, the technologies of the internal combustion engine, radios, and so on. Then the second period, right after WWII, where it's a combination of nuclear weapons, ballistic missiles, and the beginnings of computers lead to big changes. Their function, as military intellectuals, was to diagnose when there were these periods of big change. And, so they began to say one of these periods of big change was coming, because of the micro-processor and other related technologies. It was triggered, I think, by a programme to develop a system that they called the 'assault-breaker,' that conceptually was a reasonably long-ranged rocket with a smart front-end coupled to long-range sensors.

Der Derian: I was out for the first digitised rotation at Fort Irwin when the Fort Knox brigade was out there, and did some interviews. I've been looking at this from the bottom-up, from the field, and it seems there's a lot more skepticism about a revolution going on.

Marshall: Yeah, I would think so. I wouldn't particularly expect to see it down there. It's also spotty on the top, although growing, I would say. What I tend to argue with people is that we ought to see ourselves as in something like in the early twenties where we don't fully see what the outcome would be. But there is just enough, on the one hand, to see that the technologies are moving rapidly and it's plausible that there would be a big impact. We are about in a position, where people say, at the Naval War College, were about in '22 or '23, where we now have a bunch of war games that are being played, that are beginning to explore, in some sense, the logic of the situation that would exist if you had, let's say, twenty years from now, a number of new kinds of systems.

Der Derian: Are you familiar with the exercises on the Salisbury Plain in the twenties?

Marshall: Yes, '27 and so on, oh yes ...

Der Derian: Last fall I checked out the back-issues of *The Daily Telegraph*, where Liddell Hart wrote about them, and what struck me is that he didn't really call it a revolution, or understand it as such. When you are really in the middle of it, you are least aware of it.

Marshall: Yes, well I suppose to an extent there were people in the military, in the twenties that thought of things as revolutions that were primarily associated with air. Even the Germans, who really boast of being a hell of a lot better than the British, were not consciously thinking in terms of a revolution. So, what would be unique this time, in a certain sense, I think, or more so, is that because of things that have happened since the twenties and thirties — both the historical literature that has been built up that looked at these kinds of periods and the Russians who began intellectualising about it and raising it as a kind of intellectual issue — that one of the unique things about the next twenty or thirty years, if in fact we will go through such a thing, it will be almost the first time in which it is, in a widespread sort of way, self-consciously — you know — pursued or experienced as a revolution.

Der Derian: Who do you see as our next enemy?

Marshall: Well, I'm interested in Asia mainly because of some general reasons. You look at long-range projections, that's the place where the most rapid economic growth is going to take place. Also, Asia has been dominated by the West for over 150 to 200 years and that's over. And so, exactly what Asia would be like, what kind of internal rivalry will be there is something really needs to be looked at.

Der Derian: Can you really compare our times to any other? When suddenly everything is wide open, do you think a global threat is going to emerge?

Marshall: No, I think not in my time. But if you look back into history I think you can see. ... I think the twenties was like that. The twenties turned out to be a period of illusion about what the world was going to be like. I think we are in the twenties. Both in terms of the beginning of technical change which is working out its implications, and in terms of, well, in the twenties the United States didn't really have any big immediate threat, and the forces were very small. Whether something like the thirties is before us, I don't know.

Der Derian: But to what extent do we create our enemies? Do you think it is completely a unilateral action, or do you think it's more like the whole idea of the security dilemma?

Marshall: Well I don't see that right now. You have a little of that in Asia with the growth of China and how we react against it, and to the extent that we get to be seen as, you know, the people who are intervening in this place and that place ... to the extent that we have gotten ourselves in the position of being the leader of the interventions for the UN.

Der Derian: In the way that the nuclear deterrence maintained a relative peace for some, do you think there is such a thing as a high-tech deterrent, in the way that people would see what happened in the Desert Storm and then would not want to take on the U.S.?

Marshall: I think so, I mean it's a deterrent in a sense, but it also has these other effects. I mean it deters people from taking us on in this way. But it may substantially increase incentives to go after nuclear weapons, or look for other clever ways of using the technology.

Der Derian: It's clearly part of your job here, but do you really think war is persistent; we will always have it?

Marshall: I tend to be pessimistic about it and not just because of my job. If you just look at history and human behaviour you can't be very optimistic about it ...

Der Derian: You don't want to think of war as obsolescent?

Marshall: I would tend towards that view, yes.

Closing shot, a shadow looming over the White House. Ronald Reagan, at a final press conference before the National Strategy Forum in May 1988:

> But I've often wondered what if all of us in the world discovered that we were threatened by a power from outer space — from another planet. Wouldn't we all of a sudden find that we didn't have any differences between us at all — we were all human beings, citizens of the world — wouldn't we come together to fight that particular threat?

Paul Virilio, from his *Popular Defence and Ecological Struggles*:

> Meanwhile, damage and disaster occur just like the emergence of war populations, nihilistic spectre of the speed of those with no name, who cannot be named and who nonetheless arrive, those whom Chaucer in the fourteenth century already called "*builders of smokejails* ... greenish men, couriers of the Great Fear. ..." The modern myth of UFO's is mixed with that of the terrifying immanence of assassination attempts, cataclysms, crime, epidemics, enemy threats.

And finally, Walter Benjamin, from *Charles Baudelaire: A Lyric Poet in the Era of High Capitalism*:

> In times of terror, when everyone is something of a conspirator, everybody will be in a situation where he has to play detective.

Agent Mulder would probably agree.

Chapter Ten

Arms Control and the Future of International Security

Brad Roberts*

Arms control seems to have lost its sense of purpose and direction, at least in the United States. Many believe that arms control is a Cold War-vintage anachronism. Without strong political leadership from the executive branch of government and without a modicum of bipartisanship in the legislative branch, arms control has fallen prey to an old ideological debate between those who really care about arms control — its die-hard opponents and its passionate advocates. Today in Washington the debate over arms control is again dominated by those who see it as a dangerous delusion, a 'sell-out' of the national interest, and the epitome of 'do-goodism', and those at the other end of the spectrum who believe arms control is the only viable means of walking us back from the terrible consequences of a world of unrestrained proliferation. The combination has proven nearly fatal to the old moderate consensus that saw arms control as a useful instrument of national strategy when carefully conceived and tightly constructed.

In Washington, arms control is hostage not least to the passing of an agreed understanding of the role of the United States in the world. Containment of the Soviet threat provided a clear and straightforward understanding of U.S. grand strategy and a way to weigh the utility of various tools of policy in pursuit of that goal. Arms control's promise was always clear, even as the utility of specific instruments was sometimes hotly debated. It seems likely that U.S. leaders will continue to invest arms control with diminished political capital so long as its utility and promise vis-a-vis new problems of national and international security remain clouded.

It is time then to return to basics. What role can arms control play in meeting the security challenges of the emerging strategic era? If it has a role to play, must it change in order to do so? In what ways? This chapter offers

---

* This chapter is an edited version of a presentation made to the Fourteenth Annual Ottawa NACD Verification Symposium, "Cyberspace and Outer Space: Transitional Challenges for Multilateral Verification in the 21st Century," 12–15 March 1997, Montebello, Quebec, Canada.

some speculative thinking on these questions.[1] It is conceived as a brainstorming tool, aimed at assisting the reader in rethinking inherited assumptions about arms control and in beginning to evaluate some of the new dynamics of international security. It begins with a discussion of the emerging strategic era and the specific challenges of promoting orderly relations among states. It then evaluates the utility of arms control for six specific world order tasks; this is a comparative evaluation that explores a broad set of policy instruments in order to place the potential role of arms control into perspective. Finally, it explores some implications for the changing roles and functions of arms control.

World Order Tasks

Analysts have developed a number of tools for speculating about the emerging strategic era. Some look for analogies in the past, such as the decaying order of the 1920s or the swiftly rebuilt order of the late-1940s. Others construct pictures of the future by projecting forward current trends, and thus see a future dominated by transnational forces or subnational conflict. Others emphasise the transitional nature of the current era, with great uncertainty and an element of fear about what might follow, such as a renewed military competition among the major powers. This chapter adopts a different approach: it identifies six unique features of the current international system and postulates six so-called world order tasks germane to enhancing the stabilising aspects and minimising the destabilising ones of each of those features.

The current moment in world affairs is noteworthy for a variety of reasons. For the first time in more than a century, there is no significant prospect of war among the major powers in the interstate system. Yet relations among them — and factors within them — are changing, and other major powers are emerging. Maintaining this stability and reaping its benefits for the rest of the system is thus the first world order task.

At the same time, however, there is an unprecedented emergence of prosperous, developing, and increasingly powerful and ambitious countries in many regions of the world, which aspire to a status commensurate to their weight. The need to meet their expectations and to reinforce their stake in existing international institutions and norms is unprecedented. At issue is their 'alignment' and the possibility that they might become revisionist powers. Integrating these states into the existing system is thus the second world order task.

But this is also an era when localized conflicts among minor powers, or even within states, can threaten to have broad international repercussions. The advent of international terrorism and the proliferation of long-range delivery

1. Those remarks were derived from an unpublished article entitled "Arms Control in the Emerging Strategic Environment."

systems underscores this. Insulating the state system from the military competitions of small states and from weak and collapsing states is thus the third world order task.

Interstate aggression is an old problem of international relations but in a particular new guise. Successful aggressions threaten a principle that slowly has taken hold in international politics — that war is permissible only in self-defence. Many small and medium countries bank on this principle by not maintaining the strongest military forces within their reach and counting on the international community to come to their aid in time of need. Moreover, wars of aggression won by threats or the actual use of nuclear, biological, or chemical weapons could have a terrible impact on the proliferation problem, especially if the power that has backed down or been defeated is an international security guarantor. Deterring aggression and punishing transgressors of agreed norms is thus the fourth world order task.

The current international system is also noteworthy for the rapid globalisation of the world economy and for the associated boom in the exchange of technologies, materials, and expertise that are dual-use and thus militarily sensitive. Preventing or stopping the flows of dual-use items is impossible, especially given the evergrowing list of countries that generate those items. But preventing the proliferation of technologies from precipitating a proliferation of high-leverage weaponry, while also promoting accepted and legitimate uses of those technologies, is a more reasonable goal. Managing technology diffusion is thus the fifth world order task.

Finally, the current international moment is marked by the unprecedented role of the United States. It is the only state with a global view and the only one capable of mobilising international institutions to deal with threats to the peace. Its military power is unmatched — but it is no more than first-among-equals politically and economically. Moreover, although the world order requires a constancy and purposefulness to the use of American power, the United States is alternatively assertive and neoisolationist. Anchoring American power in international processes and institutions in a way that is *sufficient* to the needs of their sustained functioning and to the needs of security and prosperity more generally is an urgent requirement. Thus the final world order task is engaging the United States.

Rating Arms Control's Utility

What role might arms control play in accomplishing these world order tasks? It should be obvious that none of these tasks can be accomplished by arms control in and of itself. Indeed, there is a wide variety of political, economic, and military instruments of policy germane to each. But when and where might arms control fit into the larger strategic equation?

In some cases, arms control's contribution is likely to be, at best, modest. For integrating aspiring powers, for example, the primary tools of policy are bilateral relations with each of the emerging powers and their enhanced economic integration with the global economy. But arms control offers some specific roles and opportunities for these powers, not available elsewhere. Arms control negotiations with them on, for example, the global treaty regimes, are a way to build consensus with the major powers on the allocation of rights, responsibility, and power in the international system. Arms control is also a tool for institutionalising transparency within those societies, thus laying the foundations for their future, deeper integration into the global system.

In other cases, arms control's potential contribution seems likely to be more substantial. For insulating the state system, recent years have brought a recognition of arms control's utility in stemming the in-bound flow of small arms and the out-bound flow of weaponry and materials. It also has found a limited role in disarming factions in local conflicts — in El Salvador, for example, where such controls have been applied as a part of an agreement that also arms the legitimate police functionaries of the state. Arms control also helps to ameliorate the problem of terrorist use of nuclear, biological, or chemical weaponry or materials by criminalising certain types of activities and delegitimising the use of such weapons by states. Historically, many terrorists have been unwilling to employ weapons that states consider illegitimate, because such use might alienate their perceived constituencies and delegitimise their cause.

For deterring aggression and punishing transgressors, the primary instrument of policy is, of course, the military instrument and associated alliances and coalitions. But successful deterrence requires not just an operational capability to defeat aggression on the battlefield, but a political context that delegitimises the aggressor's actions and legitimises international responses. Arms control seems useful as a tool for building consensus about what makes some states 'rogues' — by their self-selection outside of treaty regimes or their failure to comply if inside them — for utilising that consensus to assemble political coalitions against specific states, and for collective military reply to particularly egregious illegal or illicit behaviour.

For engaging the United States, arms control has a utility generally not appreciated by those in Washington. It defines commonly agreed roles for the United States, legitimises the use of U.S. military power against regional challengers on behalf of the common interests of the international community, and provides a rule-based, institutional basis for U.S. engagement — a basis for which the United States has shown an historical affinity.

Arms control also has an important potential utility for maintaining stable relations among major powers. But its actual utility is uncertain. It seems to be an essential tool for this purpose, by structuring the most volatile strategic relationships; by facilitating dialogue amongst them about changing roles, responsibilities, and ambitions; and, by harnessing their political and military power to certain common purposes. But arms control is certainly not the primary tool for any of these purposes. Moreover, arms control poorly conceived, which is to say arms control inconsistent with the interests of specific major powers or weak arms control exploited by one party to capture strategic advantage over the other(s), could prove serious destabilising.

Implications for Arms Control

If arms control is to make its contribution to these tasks, how must it evolve? Part of the answer has to do with the function of arms control. The three classic functions of arms control, as defined in the 1950s, were to reduce the likelihood of war, to reduce its scope and violence if it were to occur, and to reduce the costs of being prepared to wage it. The foregoing review of arms control utilities suggests some different ways of thinking about the function of arms control. On the one hand, it is a process of political dialogue. In the era ahead, it would appear that this process is increasingly important, not least to those who would engage the United States in a discussion of its roles and responsibilities. On the other hand, it is a set of products whose implementation enhances stability by minimising risks. In the era ahead, it would appear that the utility of these products will be not so much remedial as preventive. In the Cold War, after all, arms control's primary appeal was in helping the superpowers to walk back from the brink of armageddon. But is it not better not to walk to the brink in the first place by formalising patterns of restraint among nations increasingly capable of mobilising domestic resources for strategic war?

Part of the answer also has to do with utilising the existing set of arms control products for new purposes. For example, the global treaty regimes for the control of nuclear, biological, and chemical weapons are useful not just for the restraints they impose on the major powers, but also for their restraints on aspiring powers, the obligation to conduct sensitive trade under conditions of transparency and control, and the dialogue they permit among major and minor powers about the proliferation threat and the security guarantor roles of the major powers. Regional mechanisms, ad hoc and informal measures, and even unilateral restraints are familiar arms control instruments with new roles and possibilities in the years ahead. New mechanisms seem likely in certain areas, such as landmines.

This analysis points to a broader set of implications for the changing arms control agenda. First, non-nuclear matters are conspicuous on the emerging agenda. Although the arms control community in the United States remains heavily focused on nuclear matters, chemical, biological, and conventional weapons — especially small arms in local conflicts — are having a growing impact on international security and thus on the arms control agenda. Unfortunately, ways of thinking about arms control, as derived from the nuclear experience, have generally obstructed a clear assessment of the distinct arms control requirements in each of these areas. For example, the verification and compliance requirements of treaties outside of the strategic domain, where cheating does not pose risks of armageddon, are not well understood.

Second, this agenda is more multilateral and less bilateral than ever. But the new prominence of multilateralism poses a special challenge to the United States, given its past focus on bilateral measures with its superpower peer and its difficult adjustment to its role as first among equals in such fora. The absence of a sustained U.S. political commitment to multilateral institutions and processes more generally has been harmful to both the Nuclear Non-Proliferation Treaty and the Chemical Weapons Convention, for example.

Third, the arms control agenda is expanding to include not just negotiation but also implementation issues. Of course, this is a question of the relative prominence in the mix, as implementation has always been a part of the arms control agenda. Implementation poses technical, organisational, and fiscal challenges that seem likely to grow in an era of weakening political commitment to arms control and sharper budgetary constraints. It also poses political challenges related to gaining full compliance by unwilling parties.

Fourth, arms control's normative content merits greater political attention. If arms control is useful in significant part as a political tool for building coalitions and for coping with technology diffusion, its political value derives from the shared norms of behaviour it reflects. But such norms are under challenge today, and not just from 'rogue' states at the margins of international politics. The perception in some quarters, that the United States pays only lip service to such norms and eschews the responsibilities of leading the institutions created to enforce them, only magnifies the problem.

Fifth, arms control's changing agenda raises fundamental questions about nuclear disarmament. For the purposes of maintaining stable relations among the major powers, deterring aggressors armed with weapons of mass destruction, and keeping the United States engaged, nuclear abolition seems likely to be counterproductive, at least for the foreseeable future. But abolition is just what is expected of the nuclear weapon states by those aspiring powers actively engaged in the NPT process — and urgently so. Moreover, the conservatism of the nuclear weapons states in addressing these nuclear questions is under growing assault from both civilians and military personnel within some of those countries. An increasingly sharp debate about the role of nuclear weapons in international security and the speed and direction of nuclear arms control certainly seems in the offing.

Sixth, the 'leadership needs' of the new agenda are poorly conceived. Leadership seems to be missing on a number of essential items: in the debate on the future of the NPT; in implementation of the CWC; in the use of the UN Arms Register and Wassenaar Arrangement to promote regional stability; in the North-South dialogue on export controls; in the dialogue with the smaller nuclear weapons states; and in confronting the critics of treaty norms, for example. But the leadership issue is not just *where* to lead — it is also *how* to lead. The temptation in the United States, at least, is to lead episodically and by cheerleading. But world order politics require a leadership style that matches vision, constancy, intellectual rigor, responsibility, and an ability to motivate others by appealing to shared interests as well as shared values.

Finally, the new agenda poses a large set of novel analytical questions. Yet arms control research has grown moribund. Thinking about arms control has become highly specialised and technical, when what is needed is a capacity to examine old instruments in new ways and to conceive new instruments for new purposes and in a new political context.

Conclusions

If there were no inherited arms control agenda, the international community would find it useful to create one for the strategic requirements of the new era. Arms control's potential contributions to national security and international stability seem to be expanding, not contracting, with the end of the Cold War.

But the reemergence of arms control as a major instrument of international action and of policies in Washington, cannot be taken for granted. Momentum alone is insufficient. Today, for example, seven arms control measures await ratification by the United States, with little prospect of such in the near term. A willingness on the part of senior leaders in Washington to invest political capital in arms control will not be easily rebuilt in the absence of a broadly accepted grand strategy for the United States and the West.

In the absence of concerted and effective leadership by Washington, other countries will have an opportunity to play new leadership roles. This is for the best. It provides new opportunities for the aspiring powers to take responsibility for larger international problems. It provides new roles for the emerging major powers like Germany and Japan. And it provides a way to compel Russia and China to examine their stake in the stability that existing international institutions are intended to serve. But if the new leadership team is unable to turn a commitment to arms control into the practice of arms control, other partners in the process will begin to opt out. International stability will suffer. It will be difficult, if not impossible, to turn around the disintegrating process in the absence of a major international crisis. And if weak or inadequate arms control is seen to have played a hand in precipitating that crisis, arms control may quickly be relegated to an historical footnote.

Chapter Eleven

Verification: An Active Role for the United Nations

Alan Crawford*

Introduction

The United Nations enters its fifty-second year of existence in 1997 — a remarkable achievement of longevity in the realm of international affairs. Yet the future of the organisation remains clouded. It faces, for example, even more turmoil in terms of financing and restructuring over the next few years. Perhaps even more disconcerting is the lack of a shared vision among Member States concerning the purpose and role of the organisation with respect to security affairs in the twenty-first century. This state of uncertainty applies doubly so when considering the role of the United Nations in the field of verification "in all its aspects."

This chapter will argue that there is a legitimate and practical role for the United Nations in certain specific and carefully delimited areas of verification. After describing briefly some of the background to the United Nations' role in verification and Canada's involvement with it, the chapter will outline what that role seems to be today. It will then move to the more problematic question of the prospects for supporting and enhancing this role, both in terms of desirability and feasibility. Finally, it will address what Canada should and can do in this respect.

Background

Canada has, for a number of years, supported the consideration of an active role for the United Nations in verification, as indicated in **Table 1**. Beginning in 1985, when verification remained a contentious East-West issue, Canada initiated the first United Nations General Assembly (UNGA) resolution that specifically addressed the subject.[1] That resolution articulated the view that verification of arms control and disarmament (ACD) agreements was important. That initial resolution has been followed, at first annually and later biennially, by further UNGA resolutions on which Canada has taken the lead. In the initial years, these

---

* This chapter is an edited version of a presentation made to the Fourteenth Annual Ottawa NACD Verification Symposium, "Cyberspace and Outer Space: Transitional Challenges for Multilateral Verification in the 21st Century," 12–15 March 1997, Montebello, Quebec, Canada. The views expressed in this chapter are those of the author alone and do not necessarily represent those of the Department of Foreign Affairs and International Trade or of the Government of Canada.

1. United Nations General Assembly, Resolution 40/152 O, 16 December 1985.

resolutions were adopted without a vote, reflecting an international consensus. More recently, this consensus has broken down.

In 1987, Canada took the lead in initiating a United Nations Disarmament Commission (UNDC) working group and chairing that group.[2] The UNDC eventually produced, in 1988 by consensus, a report containing, among other things, a significant set of sixteen "verification principles."[3] That same year, the UNDC exercise was followed by an UNGA resolution that set up a United Nations Group of Governmental Experts Study to examine the role of the United Nations in verification.[4] That resolution saw the first rift in the international consensus on the issue, with the U.S. casting a single negative vote on the grounds that verification was a treaty-specific exercise and a United Nations role in verification could not be examined in the abstract.

The United Nations Study Group, under the Chairmanship of Canada's Fred Bild and comprising governmental experts from twenty countries, including a U.S. participant, still confronted much of the political 'old-think' of the dying days of the Cold War. Nevertheless, it managed to produce by consensus, an interesting report in 1990.[5] The 1990 Report discussed the concept and functions of verification, described existing United Nations activities in the field, and articulated a few very modest ways to enhance the role of the organisation relating to construction of a verification database and promoting the exchange of views between experts and diplomats. The Report noted that the development of a United Nations "verification system," a point which was promoted by several Third World members of the Group, depended on further changes in the political environment and the evolution of arms control agreements, concluding only that consideration of such a verification system should continue. On balance, given the constraints of the Cold War, this report achieved a remarkable consensus between East and West, as well as between North and South. The 1990 Study was adopted in UNGA Resolution 45/65 of 4 December 1990, which was agreed without a vote, suggesting that international consensus had been re-established.

Later, however, international consensus on the issue within the United Nations began to deteriorate again. In the aftermath of the 1991 Gulf War, the Security Council had set up the United Nations Special Commission (UNSCOM), which was tasked, along with the IAEA, to verify the disarmament of Iraq's weapons of mass destruction and long-range missiles.[6] UNSCOM and the IAEA were given rights to inspect and monitor Iraqi activities that were unprecedented in their comprehensiveness and intrusiveness. Other international events had also

---

2. United Nations General Assembly, Resolution 42/42 M, 30 November 1987.
3. *Official Records of the General Assembly, Fifteenth Special Session, Supplement No. 3* (A/S-15/3).
4. United Nations General Assembly, Resolution 43/81 B, 7 December 1988.
5. *Study on the Role of the United Nations in the Field of Verification*, A/45/372, 28 August 1990.
6. United Nations Security Council, Resolution 687, 3 April 1991.

unfolded since 1990 that had significant meaning for multilateral verification, including the adoption on 19 November 1990 of the Conventional Forces in Europe (CFE) Treaty.

In 1992, the United Nations Secretary General produced *An Agenda for Peace*,[7] which was followed by several other documents dealing more explicitly with ACD issues.[8] These documents possessed a tone, understandable perhaps in the aftermath of the successful remedial action by the international community to expel Iraqi forces from Kuwait, which seemed to favour a much more activist and interventionist role for the United Nations. The United Nations, it seemed, would no longer stand passively by to allow developments within a country to grow beyond its borders into a threat to international peace and security. The result was a growing concern among Third World states that the United Nations might increasingly interfere in their internal affairs.

Another important trend, increasingly apparent after 1990, was the success by the international community in achieving progress in reaching agreements addressing weapons of mass destruction, most notably the Chemical Weapons Convention. As a result, more attention began to turn toward dealing with the problems generated by conventional weapons, particularly the accumulation of excessive and destabilising quantities of these weapons. This was perhaps best exemplified by the creation of the United Nations Register of Conventional Arms. Also, there was increasing focus on the impact of such conventional weapons, including light weapons in local conflicts, especially where peacekeeping forces might be operating, and on humanitarian concerns generated by the mis-use of specific conventional weapons such as anti-personnel landmines. Increasingly, attention seemed to be shifting away from weapons that were largely a monopoly of the big powers — i.e., weapons of mass destruction — to those used by smaller and poorer countries — i.e., conventional weapons. The result was a heightened sense of unease on the part of the latter's governments.

At the same time, many of the big powers increasingly questioned any enhanced security role for the United Nations, including in the field of non-proliferation, arms control, and disarmament (NACD). In part, this reflected dissatisfaction with the perceived inability of the UN to work effectively in this area and with the 'messiness' of United Nations politics. In addition, it may well reflect an antipathy toward creating a strong supra-national organisation that would have the power to constrain the big powers' freedom of action. As a result, the U.S. and some other states displayed an increasing hostility to studying an effective United Nations role in verification, or even further consideration of the subject.

Finally, the mid-1990s saw the United Nations — as well as its main donors — facing an increased financial crisis. There was less money to support an enhanced

---

7. Boutros Boutros Ghali, *An Agenda for Peace*, A/47/277, 17 June 1992.

8. Boutros Boutros Ghali, "New Dimensions of Arms Regulation and Disarmament in the Post-Cold War Era," A/C.1/47/7, 23 October 1992; and *idem*, *Supplement to an Agenda for Peace*, A/50/60, 3 January 1995.

role for the UN in security affairs, let alone verification. Indeed, given the greater number of local conflicts to which the United Nations was being called upon to send peace operations, there was growing concern that the UN would not be able to perform its traditional security functions, much less expand its responsibilities.

Against this background, in 1994, Canada initiated another United Nations Group of Experts Study to explore a UN role in verification. The rationale was that the international environment had changed greatly since the end of the Cold War, at the end of which the 1990 Group had conducted its work, and that verification of multilateral NACD obligations was becoming increasingly important. The establishment of this new Group of Experts Study was resisted by many of Canada's traditional allies, including the U.S. and European Union (EU) states, primarily on the grounds that it was too soon, examining UNSCOM's experience was too sensitive, and it would be too expensive. From a purely research perspective, all three excuses were suspect. Through perseverance, the Group was established by UNGA Resolution 48/68 of 16 December 1993. Under the Chairship of Canada's Peggy Mason and with only a Dutch and Swedish expert from other developed countries, the Group reached a consensus report in 1995.[9] That report was not, however, adopted by consensus in the General Assembly: the United States, Iraq, and North Korea, among others, did not support it.[10] In a welcome gesture, Germany, after abstaining on the resolution in the First Committee, changed its vote to an affirmative in the Plenary of UNGA, after having had an opportunity to evaluate the report more carefully.

The 1995 United Nations Study introduced an important conceptual refinement to the concept of verification: it expanded the application of the concept beyond formal NACD "agreements" to encompass "obligations," reflecting, in part, thinking that was present in previous UNGA resolutions concerning compliance with NACD obligations. This expanded concept of verification now applied to obligations imposed by the United Nations Security Council (such as in Iraq), including arms embargoes, and to less formal agreements as might occur in the context of the operation's peacekeeping missions. The 1995 Study also outlined a more ambitious — though still modest and very qualified — set of recommendations for an enhanced United Nations role in verification. These are summarised in **Table 2**.

Verification Activities of the United Nations Today

In theory, the United Nations has a major role in international security deriving from its Charter — in practice, this role has been and still is subject to major restrictions imposed by Member States, especially by the big powers. However, the very nature of the international system and its focus on sovereign states also constrains UN activities in international security and, by extension, in verifica-

9. *Verification in All Its Aspects, Including the Role of the United Nations in the Field of Verification*, A/50/377, 22 September 1995.

10. United Nations General Assembly, Resolution 50/61, 12 December 1995.

tion. Traditional concepts of national sovereignty favoured by both big and small powers limit the role the United Nations can play.

Norm-Building

One role about which there appears to be a fair degree of international consensus and one where the United Nations has made some important contributions is in the development and articulation of international norms relating to verification. The UNDC sixteen principles mentioned above are an example of this. One might also hope that the expanded definition of verification outlined in the 1995 United Nations Study would also prove to be an example, though this remains to be seen.

This is an important role, though one that is often underrated. Despite fatigue on the part of many with rhetorical pronouncements or the 'lowest common denominator' character of many United Nations documents, most people would recognise the need for some body that can make authoritative statements that represent the consensus, or at least majority beliefs, of the international community. In the context of verification, the UN has succeeded in increasing international recognition of the importance of verifying compliance with NACD agreements and obligations: relying on good faith declarations is simply not enough when important issues relating to security are at stake. The need to balance the requirement for verification with concerns about national sovereignty and other legitimate matters is an example of an important general principle about which there is also broad international agreement. The United Nations can claim considerable credit for helping to develop, articulate, and maintain international consensus about such ideas.

However, norms are not built and do not endure by grandiose statements of principles alone. They are generated and gain much of their meaning by the practice of Member States. The United Nations has had less success in contributing directly in this area.

Verification Infrastructure

Both the 1990 and the 1995 United Nations studies suggested that the UN can play a useful role in contributing to the development of what might be called a "verification infrastructure," by which is meant certain shared management and coordination functions on behalf of the international community. These functions are usually conceived in much less treaty-specific terms, in contrast, for example, to the tasks that the Verification Implementation and Coordination Staff of NATO performs for the Parties to the CFE Treaty. Instead, the United Nations' role is seen as applying generically across treaty regimes. Among the functions suggested are support for verification research, training, and the exchange of views/communication between treaty regimes and between Member States. Because of its universal membership, the UN is thought by some to be in a better position to see the broader picture and to be better able to exploit potential synergies that might not be apparent to a treaty-specific organisation or to individual Member States.

Such a role for the United Nations may be useful in certain narrow contexts and may be very important to some Member States, particularly those lacking national capabilities. Success in this role, seems to be a matter of the UN carefully identifying and focussing on specific requirements in this area. However, it must be recognised that the United Nations is not alone in undertaking such "infrastructure" activities. Non-governmental organisations, including research institutes and also some Member States, also make important contributions. It is debatable, therefore whether the UN really has a *unique* role to play in developing an international verification infrastructure at this time.

Operational Activities

The real 'meat and potatoes' of NACD verification is in operations. It is in this area that the UN's role is, ironically, both very restricted and almost pre-eminent.

It is useful at this point to digress briefly, to outline some of the advantages that have been advocated in the past for a comprehensive international verification organisation — that is, one that would undertake verification of a variety of NACD agreements. These advantages might also suggest a more robust role for the United Nations in operational verification activities. Among these advantages are the following:

- verification by such an international body can achieve economies of scale and thus be cheaper than by groups or individual countries;
- for smaller countries in particular, which lack national capabilities, an international body can help provide a vehicle for meaningful participation in verification activities;
- important synergies and cross-fertilisation can be achieved because of the international body's broader scope of coverage, with the result that it can verify compliance more effectively than can a number of treaty-specific organisations or individual countries; and,
- such a role for an international body helps promote the rule of law and international governance.

Returning to the question of the United Nations' current verification role in the realm of traditional NACD, by which is meant verification related to specific formal NACD agreements, the UN has little operational responsibility. There are two exceptions to this general statement. First, the United Nations has undertaken and still does perform some fact-finding activities with respect to compliance with the 1925 Geneva Protocol dealing with chemical and biological weapons use. This function is expected to be largely taken over by the Organization for the Prohibition of Chemical Weapons (OPCW) when the Chemical Weapons Convention comes into force. Second, if one considers the International Atomic Energy Agency (IAEA) to be part of the United Nations, then the UN does have a central role in verifying the Nuclear Non-Proliferation Treaty (NPT). However, this is, at best, a quasi-exception, because the IAEA operates, for the most part, quite independently of the central United Nations organs.

There has been a marked lack of enthusiasm among the parties to formal multilateral arms control agreements for giving the United Nations a role in the verification of those agreements. In fact, it is hard to avoid the impression that parties have gone out of their way to prevent UN involvement. Why is this the case? Perhaps there are several reasons: disillusionment with the perceived inefficiency and ineffectiveness of the United Nations; fear of strengthening the United Nations' power and promoting a supra-national government; and, the desire to ensure the requisite technical expertise. Another important consideration is the fact that each separate arms control agreement is composed of a different set of parties — sets which do not coincide with the membership of the United Nations. In other words, giving the UN authority with respect to verification of a specific NACD agreement would mean allowing non-parties to that agreement to have a say in its verification. A valid point; however, it has not prevented the IAEA from being assigned responsibility for NPT verification. The membership of the IAEA includes non-parties to the NPT.

Also, Third World countries, as mentioned above, may fear that an activist United Nations might be used to interfere with their internal affairs and weaken the traditional concept of national sovereignty. Finally, some may be concerned about the financial implications in an era of tight United Nations budgets. It is also worth mentioning that there has been no interest on the part of Russia/USSR and the United States for a United Nations role in verifying their bilateral NACD agreements. Clearly, the question of superpower status and power are key in this context.

This situation with regard to formal traditional NACD agreements contrasts sharply with the more informal NACD verification in peace and security operations, by which is meant United Nations peacekeeping, peace building, peace enforcement, etc. The 1995 United Nations Experts Study pointed out that peace operations often involve an important NACD component and that the UN has a great deal of experience in this area. More recently, the term "micro-disarmament" has been introduced to describe this activity.

For peace operations, neutral third party verification is a key idea — that is, involvement by a trusted third party which undertakes a verification role in lieu of or in addition to activities by the adversaries. Of course, there are peace operations where the United Nations has no role — for example, the Dayton Accords where the OSCE is playing the role of third party verifier in some ways. Another example is that of the Multilateral Force and Observers with respect to the Egypt-Israel Peace Treaty. Moreover, there has been at least one important instance wherein this third party verification role was undertaken by an individual country: the U.S. during the Sinai Disengagement process between Egypt and Israel. Nevertheless, the UN can rightfully claim pre-eminent experience in this area.

A unique case of United Nations involvement in verification is that of UNSCOM/IAEA in Iraq. Clearly, this is not a situation of verifying a traditional NACD agreement, or even of United Nations micro-disarmament activities. For a number of reasons, this case is special: it involves a unique United Nations Security Council mandate; the weapons involved are weapons of mass destruction, unlike the case in micro-disarmament; it is decidedly compulsory in nature; and, the degree of intrusiveness and intensity of the disarmament involved is unprecedented. Nevertheless, the ongoing experience in Iraq has much to teach about the methodology of verification — indeed, the 1995 United Nations Study described it as a "verification laboratory."

Another area where the United Nations has extensive involvement is one that is often overlooked: that of arms embargoes. Because the *monitoring/verification* element in embargoes is intimately intermingled with the *enforcement* of the embargoes, and because arms embargoes are also intertwined with more general economic embargoes, such cases have special characteristics. Nevertheless, as the 1995 United Nations Study recognises, arms embargoes are a form of arms control in the sense of having an obligation relating to the acquisition of weapons and monitoring compliance with that obligation. Monitoring and verification in this area is most often *ad hoc*, rudimentary, and inefficient — it is left up to Member States to implement, for the most part. There are many political and technical complications — especially demands for the compensation of economic losses by innocent parties and the involvement of customs controls. Verifying arms embargoes is also a costly activity. As a result, there has been little political will to improve the capability of the United Nations in this regard, and little research has been done in this area.

One great difficulty in describing the foregoing activities by the United Nations — peace operations, UNSCOM, arms embargoes — as NACD verification is that the peacekeeping and diplomatic community tend not to think of them as involving arms control, verification, or even monitoring. This is largely a result of a preoccupation with formally negotiated agreements and a question of terminology, as well as, perhaps, bureaucratic mindsets. All the basic functional elements of NACD verification seem to be present in micro-disarmament, the UNSCOM experience, and monitoring arms embargoes, despite their special individual characteristics.

The foregoing suggests that, despite the traditional wisdom on the subject, the United Nations today has an important role in certain types of NACD monitoring and verification. This is not the old idea of the UN as a supra-national verification organisation, but it is nevertheless a very important role.

Prospects for Enhancing the United Nations Role in Verification

Norm-Building

Verification is on the agenda for the next United Nations General Assembly in October 1997. However, while the UN still has a role to play in promoting international consensus about verification of NACD obligations, there is decidedly little interest among Member States in pursuing this agenda item at this time. Most do not see it as a priority. Most, while paying 'lip-service' to verification of traditional NACD agreements, do not really see it as critical and, for reasons outlined above, are unlikely to see the United Nations as having a central role in this area. As well, most Member States have trouble seeing the NACD element in peace operations and, if they do accept this conceptual leap, see peace operations as being so different from traditional NACD that there is nothing to be learned from one to the other.

Apart from a modest resolution reiterating the importance of verification, it is difficult to see where the UNGA can go on this matter. Certainly, any reference to the 1995 United Nations Study will cause a break in consensus, as occurred in 1995. Proposing further United Nations consideration of verification principles and guidelines is also unlikely to be met with much enthusiasm because this is not seen as a priority. One possibility might be to try to push for greater recognition of the United Nations' role in monitoring and verifying the NACD element in peace operations and arms embargoes. However, a great deal of political baggage will have to be overcome if such an effort is to be successful.

In terms of norm-building, a more important effort might be to focus on regional organisations such as the ASEAN Regional Forum (ARF) and the Organization of American States (OAS). Such a regional focus would concentrate on emphasising the important role that NACD can play in regional security affairs and the critical nature of effective verification for successful NACD.

Verification Infrastructure

While improvements in this area might be a useful task for the United Nations to undertake, it is hard to justify them as essential or a priority, especially in the environment of severe resource constraint faced by the organisation. If anything is to be done in this area, it will likely have to be financed by special efforts of Member States. Despite the recommendations of the 1990 and 1995 United Nations studies, the organisation has done little to develop a United Nations verification database. One possible action might be for Canada to press ahead, perhaps with United Nations Institute for Disarmament Research (UNIDIR) assistance, to create such a database.

Another idea might be to promote a "verification fellowship" or training programme especially intended for Third World diplomats and officials. This might best be done with the support of a consortium of Member States following the model of the United Nations Disarmament Fellowship.

Finally, thought might be given to establishing a regular process or forum under the rubric of the United Nations for promoting the exchange of views among various NACD verification regimes. The easiest thing to do in this regard might be to establish an annual, or perhaps biennial, conference where representatives from various treaty bodies — i.e., IAEA, OPCW, CTBT Organisation — peace operators — i.e., UNSCOM, UN/Department of Peacekeeping Operations — regional organisations — i.e., Organization for Security and Cooperation in Europe, NATO/Verification Implementation and Coordination Staff — as well as others involved in verification/monitoring, could meet to share informally their views on operations and concepts.

Needless to say, any of these undertakings would not be cheap. On the other hand, we are not talking about an outrageous expense. Moreover, such modest initiatives might have a real benefit in terms of promoting NACD verification expertise, cross-fertilisation of ideas, and building an international verification infrastructure.

Operational Activities

In the realm of traditional NACD, apart from the special case of the IAEA, there seems very little likelihood for enhancing the role of the United Nations in verification. Most Member States seem very wary of this idea: the big powers because they don't want to give the United Nations real independent power over their own activities; and smaller less-developed states because they fear an interventionist United Nations that would be used by the big powers against their interests. Moreover, there is a natural reticence by the parties to formal NACD agreements to allow non-parties (under the guise of the UN) to have a say in verification of those agreements; the United Nations is not seen as a neutral third party, as it might be by adversaries in peace operation contexts. There is also some logic to the contention that verification operations are essentially "agreement specific," because it is the obligations of a specific agreement that are being verified, perhaps involving unique methodologies. On the other hand, does this really mean that one organisation cannot verify obligations deriving from different agreements? Do the RCMP, FBI, or customs officials, for example, only enforce one act of legislation? Whatever the merits, the "agreement specific" argument will be used to constrain expansion of a United Nations verification role. Finally, the poor reputation of the UN for inefficiency and ineffectiveness is a severe constraint for many Member States. In sum, a United Nations-based international verification organisation covering existing NACD agreements remains a remote possibility.

One potential exception to this general statement is the United Nations' residual role in dealing with cases of non-compliance with NACD obligations that are referred to the Security Council by treaty-specific organisations or by Member States. The Security Council has already stated that it will consider proliferation of weapons of mass destruction as a threat to international peace and security.[11] One idea suggested to facilitate this UN function has been the development of a Security Council capability to investigate or verify non-compliance — an idea sometimes linked to UNSCOM or a successor body. In addition, the Secretary General already has an investigative power which has been used in the past — i.e., with respect to the 1925 Geneva Protocol case — and could be used in the future in a NACD verification context. Both the 1990 and 1995 United Nations Verification Studies called for the strengthening of this fact-finding capacity. In some limited ways this has been done through the establishment of rosters of national experts, though the impression remains that this process is still largely *ad*

11. United Nations Security Council, "Note by the President of the Security Council," S/23500, 31 January 1992, 4.

hoc in nature and not as efficient as it might be. Reliance is still placed on contributions from Member States rather than an independent UN capacity.

In a related vein, there have be suggestions for improving the capacity of the United Nations to analyse verification-like data, such as that flowing from various NACD registers like the Conventional Arms Register, or information related to conflict prevention/crisis management situations. The analysis of overhead imagery might form one dimension of such a capability. The recent DFAIT report on a United Nations Centre for Information, Training, and Analysis (CITA) is one example of such ideas.[12] Canada's proposals regarding a United Nations Rapid Reaction Capability also incorporates similar notions.[13]

As publicly available information grows in volume and technical nature, traditional United Nations analytical processes may no longer be adequate. Clearly, there is a perception by many that the UN needs to strengthen its ability to collect and analyse in a timely fashion information relevant to its peace and security functions, independent of the contributions that Member States might choose to make. Such a capability could, among other things, be used to address NACD verification questions. But this development would entail giving the UN an *independent* capacity to analyse and come to conclusions. It is far from clear that Member States want this and it seems unlikely that it will happen, at least in formal terms, in the near future, except perhaps in strictly defined *ad hoc* situations such as UNSCOM. It is possible, nevertheless, that such a capability might evolve *informally* within the United Nations simply because it is essential if the UN is to function in a meaningful way in the peace and security field.

In the realm of non-traditional NACD verification operational activities, the United Nations will remain an important player. There are probably a host of practical, if mundane, ways why the United Nations' capability to undertake NACD verification in peace operation contexts should be strengthened. The 1995 United Nations Study suggested a few. This is probably an area where significant research and action can be expected, even in these tight financial times. However, while there might be an even greater recognition of the value of such improvements, they are unlikely to be seen in NACD terms.

Future cases of peace enforcement will probably be rare, but important when they happen. Can the UN's capacity in this area be improved? Probably, but given the unpredictability and limited likelihood of such operations, it is difficult to see specific actions being taken in this regard. Any United Nations capacity to verify NACD obligations deriving from peace enforcement will more likely derive from

---

12. Patricia Bliss McFate, F. Ronald Cleminson, Sidney N. Graybeal and George R. Lindsey, *Verification in a Global Context: The Establishment and Operation of a United Nations Centre for Information, Training and Analysis (CITA)*, Arms Control Verification Studies No. 7 (Ottawa: Department of Foreign Affairs and International Trade, February 1996).
13. Canada, *Towards a Rapid Reaction Capability for the United Nations* (Ottawa: September 1995).

capacities developed for other contexts — such as peacekeeping — or be *ad hoc* in nature, as is UNSCOM.

With respect to arms embargoes — one type of peace enforcement, if you like — the United Nations will probably remain an important actor, but there seems to be little prospect for improvement of the chaotic state of embargo monitoring/verification. This area too is not usually seen as being a case of NACD verification. Nevertheless, there is a crying need to undertake serious research into this area, if only because more effective embargoes might be an effective tool for the international community short of armed intervention.

There may well be an NACD dimension to cyberspace. Whether there will be an important United Nations involvement in this context is not at all certain. A UN role in verifying NACD obligations with respect to outer space seems unlikely in the near future. In the longer-term, a more active United Nations role might be conceivable.

Canada and the United Nations' Role in Verification

It would be easy to question why we should waste our time and resources on a United Nations role in verification, especially in the face of the indifference of the international community. But NACD will continue to be an important dimension of international peace and security. The parameters and nature of some NACD obligations may be changing in some respects, but monitoring and verifying of compliance with such obligations remains vital. It is just as true today as it was during the Cold War that declarations of good behaviour are not sufficient. Moreover, increased pressures to develop innovative and cost-effective approaches to monitoring and verifying NACD obligations can be expected.

There is a Canadian bias toward improving the United Nations role and capacity in the security field in general. The UN has long been seen as central to Canadian interests, including in the area of international peace and security. This perspective is not, however, shared by all Member States, affecting what action is feasible.

Whether NACD verification is one area of the security field where the United Nations role and capacity should and can be enhanced is a difficult question to answer. An effort has been made in this chapter to outline areas where the United Nations already has a legitimate role and where it is active. In some of these areas, there are possible avenues where the UN could improve its performance; in some areas there appears to be a fair degree of agreement though not necessarily consensus that the United Nations needs to do so. With respect to verification norm-building and infrastructure functions, some activities could be pursued, which are not too expensive and which would have some positive impact. But the question remains as to just how important these functions are. The answer is probably that they are useful but not essential.

With respect to verification operations, enhancing the role of the United Nations with respect to traditional formal NACD agreements is probably a non-starter. Despite reasonable arguments that there might be real cost savings and synergies to such a role for the UN, it is not politically feasible at this time or in the foreseeable future.

Perhaps the most important area where the United Nations' role can and should be enhanced relates to NACD in peace operations. This will likely happen regardless of whether the NACD community is involved or not. A significant constraint on inputting an NACD perspective into this process is that those who are responsible for peace operations, both within the United Nations and within national bureaucracies, are different from those dealing with NACD. This different bureaucratic structure follows from traditional categories of security thinking. There is a need for better communication and integration of these processes and structures. While the need may be present to improve United Nations verification of arms embargoes, there seems little likelihood of much progress in this area, though further research activity would be beneficial.

What is the bottom line? It probably relates to whether the international community wants a norm-based international system and whether it is willing to pay the price of achieving this aim. In domestic society, monitoring and verification of compliance with norms goes on all the time, whether the issue is criminal or contractual in nature. If we want a similar norm-based international society, then verifying and monitoring compliance will also be an essential ingredient. Without this function, it is difficult to see how such a system could operate — good faith is simply not enough. To quote from the first submission to the United Nations by Canada on the question of verification:

> [Verification] should meet the need to institutionalize in the context of relations among states the kind of accepted rules, procedures and expectations as those that govern the conduct of relations among individuals in all civilized societies. Such rules and procedures do not presume bad faith or malevolent intent on the part of others, but they allow for such a possibility and provide a framework in which unjustified accusations could be authoritatively rebutted, misunderstandings clarified and resolved, and non-compliance objectively established.[14]

To the extent that the United Nations is a key player in the peace and security dimension of the international system, and it seems likely to continue to be so, then the organisation should have an active role in certain carefully defined areas of verification.

---

14. Canada, [Response to Resolution 40/152 O], in United Nations Secretary General, "Verification in All Its Aspects: Report of the Secretary General," A/41/422, 11 July 1986.

Table 1: A Brief Overview of United Nations Treatment of the Issue of Verification

Year	Name	Description	Outcome
1978	Second UNGA Special Session on Disarmament (UNSSOD II)	Included some basic verification concepts in its Final Document.	Adopted by consensus.
1985	UNGA Resolution 40/152 O	First UNGA resolution devoted to verification exclusively. Emphasised importance of verification. Initiated by Canada.	Adopted without a vote.
1986	UNGA Resolution 41/86 Q	Requested UNDC to consider the subject of verification. Initiated by Canada.	Adopted without a vote.
1987	UNGA Resolution 42/42F	Requested UNDC to continue its consideration of verification. Verification became an independent item on the agenda of the UNGA. Led by Canada.	Adopted without a vote.
1988	UNDC Working Group Report on Verification (A/S-15/3).	The UNDC report contained sixteen agreed verification principles, a section on provisions and techniques of verification, and views on the role of the UN and its Member States in verification. Canada chaired the UNDC Working Group on Verification during 1987 and 1988.	Adopted by consensus.
1988	UNGA Resolution 43/81 B	Acknowledged that the UN was already playing a useful role in verification and can make a significant contribution; endorsed the UNDC sixteen verification principles; and, requested the UN Secretary General to undertake a Group of Experts study on the role of the UN in the field of verification. Canada and Sweden led on this resolution.	Recorded vote of 130 to 1 (U.S.) with no abstentions. The U.S. stated it voted no because verification arrangements must be developed and agreed by negotiating parties and it did not see how the Secretary General could undertake a study on the role of the UN in verification in the abstract.
1989	UNSSOD III	Deliberations focussed on the question of the role of the UN in verification.	No Final Document adopted, although there seemed to be a consensus emerging on verification.

1990	UN Group of Experts Report on the Role of the UN in the Field of Verification (A/45/372)	The report analysed the concept of verification including its definition and functions; described a range of verification methods and techniques, as well as existing activities of the UN in verification; and, made three general recommendations concerning improvements to the UN role: (1) a UN data-bank on verification, (2) promotion of exchanges between experts and diplomats, and (3) support for the Secretary General's fact-finding activities concerning compliance with ACD agreements.	Report adopted by consensus.
1990	UNGA Resolution 45/65	Welcomed the Group of Experts report and requested Secretary General to take appropriate action within existing resources on the report's recommendations. Led by Sweden.	Adopted without a vote.
1992	UNGA Resolution 47/45	Requested Secretary General to seek views of Member States *inter alia* on how verification can facilitate UN activities in peace operations, and on additional actions concerning the role of the UN in verification, including further UN studies. Led by Canada.	Adopted without a vote.
1993	UNGA Resolution 48/68	Requested the Secretary General to undertake a follow-on Group of Experts study, in view of significant developments in international relations since 1990, that would examine the lessons from recent experience, with particular attention to ways verification can facilitate UN activities in confidence-building, conflict management, and disarmament; explore further development of guidelines and principles; and, review the conclusions of the 1990 Study. Led by Canada.	Adopted by 145 to 0 with 22 abstentions (incl. EU members and U.S.). In explaining its abstention, the U.S. said it was premature to undertake a new study; it was too sensitive to examine work of UNSCOM and UNSCOM experience had doubtful relevance to arms control and other UN activities; and, the expense of a new study should be avoided in view of the financial burden on UN of new peace operations. Belgium, on behalf of EU members, explained their abstentions in similar terms and added that it was inappropriate to deal with verification in the abstract outside treaty specific contexts.

Table 1 (cont'd)

Year	Name	Description	Outcome
1995	UN Group of Experts Report on Verification in All Its Aspects, Including the Role of the UN in the Field of Verification (A/50/377)	The report expanded the 1990 Study's definition of verification beyond formal agreements to include "obligations"; reviewed the conclusions of the 1990 Study and implementation of its recommendations; described UN verification and other relevant international developments since 1990; explored possible future activities by the UN in verification; and, made recommendations concerning the role of the UN in verification in three general categories: (1) facilitation/coordination, (2) common services, and (3) operations.	Adopted by consensus.
1995	UNGA Resolution 50/61	Took note of the 1995 Group of Experts report and commended it to attention of Member States, and encouraged Member States to assist the Secretary General in implementing its recommendations. Led by Canada.	Adopted 157 to 1 (U.S.) with 6 abstentions (DPRK, France, Georgia, Israel, Monaco, UK). In explaining its negative vote, the U.S. reiterated it had not supported setting up the study and could not endorse its conclusions or recommendations. The UK also said it could not endorse the study's recommendations which ran counter to the view that verification measures should be treaty-specific.

Table 2: A Synopsis of the Results of the 1995 United Nations Groups of Experts Study

Ideas for Verification Guidelines and Principles:

- the importance that transparency has for verification;
- the importance of verification for providing early warning of non-compliance with obligations;
- the role that neutral third party verification can play, particularly when there is hostility among the parties to a conflict;
- the need to verify the absence of *un*declared activities inconsistent with obligations, as well as verifying declared information;
- the growing importance of cost-effectiveness in verification as well as of using the synergies between verification methods and between verification organisations to improve verification;
- the value of pooling verification resources among parties and of common services by an international organisation;
- the need for suitable start-up periods and joint training and research;
- the benefits of sharing information from national and multilateral sources; and,
- the importance of independence and impartiality for international verification organisations.

Recommendations:

a) *facilitation/coordination:*
- promotion of the exchange of verification experience; and,
- encouragement of cooperative monitoring and verification experiments.

b) *common services:*
- development of a number of databases relating to verification;
- establishment of a modest, operationally-oriented centre for information, training, and analysis;
- development of standard procedures and channels for, as well as encouragement of, sharing with the UN of national source verification data;
- development of a training programme for verification implementers; and,
- expansion of existing agreed verification guidelines and principles.

c) *operations (neutral third party verification):*
- provision by the UN of verification assistance on request;
- exploration of better preparation and systematisation of verification in peace operations and in sanctions;
- systematic collection of verification experience deriving from UN peace operations; and,
- investigation of the leasing or purchase of commercial remote sensing aircraft for UN verification activities.

Chapter Twelve

Aerial Surveillance in the Sinai Field Mission, Multinational Force and Observers, and the United Nations Special Commission on Iraq: Issues and Commonalities

René Unger*

Introduction

This chapter explores the use of aerial surveillance in the Sinai Field Mission (SFM), the Multinational Force and Observers (MFO) which followed it, and the United Nations Special Commission on Iraq (UNSCOM). It will compare the aerial surveillance methods and technologies used in these cases. While the broader contexts of the SFM/MFO and UNSCOM appear quite dissimilar, this chapter suggests that there are commonalities between these two extremes of the arms control verification experience. These commonalities are explored through a discussion of such issues as cost, the degree of interaction between aerial surveillance activities and other verification efforts, the complexity of sensors required, the relative merits of high- versus low-altitude surveillance, and the procedures needed to assess information gleaned from aerial surveillance.

The Sinai Field Mission and the Multinational Force and Observers

Established under the auspices of the Sinai II Agreement and the Sinai Support Mission (SSM), the SFM was created to provide Israel, Egypt, and the United States with information regarding unauthorised movements in the strategically important Gidi and Mitla passes. Its creation represented a step in the long process of stabilising relations between Israel and Egypt in the wake of the 1973 war. The process began with the Kilometre 101 Agreement in November of 1973, in which Israeli and Egyptian military officers agreed upon a Six-Point Agreement that, among other things, saw the reintroduction of the United Nations Emergency Force (UNEF II) into the region. This was, in turn, followed by the January 1974 Sinai Separation of Forces Agreement — also known as the Sinai I Agreement — which established the first clear lines to separate the two belligerents

---

* This chapter is an edited version of a presentation made to the Fourteenth Annual Ottawa NACD Verification Symposium, "Cyberspace and Outer Space: Transitional Challenges for Multilateral Verification in the 21st Century," 12–15 March 1997, Montebello, Quebec, Canada.

as well as the groundwork for a process of disengagement. The Sinai I Agreement also created a demilitarised buffer zone controlled by UNEF II and two limited force zones to be inspected by UNEF II. The September 1975 Sinai II Agreement built upon this process with the establishment of national surveillance stations as well as the creation of the SFM and the Joint Commission. It also described aerial reconnaissance provisions for the U.S., Israel, and Egypt, as well as provisions for the monitoring of the two passes. The final Israeli withdrawal from the Sinai was achieved with the March 1979 Egypt-Israel Peace Treaty.

Both the U.S. and the UN played supportive roles as neutral third parties to the Sinai I and II Agreements and the Peace Treaty. American 'shuttle diplomacy' enabled Israeli and Egyptian negotiators to slowly transcend their initial distrust, and the promise of an American presence in the Sinai represented an invaluable tool in enabling the Israeli government to overcome domestic pressure from a number of Israeli groups attempting to scuttle the peace process. An American and UN commitment to assist in the monitoring of activities in the Sinai and to provide both Egypt and Israel with this data represented a crucial element in the disengagement process outlined in the Sinai II Agreement. Indeed, the success of the verification activities of the SFM and UNEF II should be considered within the context of the self-interest which both Israel and Egypt derived from the Peace Treaty. Through this document, the United States both insulated them from sudden shifts in the strategic character of this region and ensured the effective monitoring of this sensitive border area — witness the American promise to Israel and Egypt to "ensure the establishment and maintenance of an acceptable alternative multinational force" in the event of a UN failure to carry out the terms of the Peace Treaty. In this regard, the arms control effort in this region was initiated and strengthened by a superpower-generated balance of both security considerations and access to information between the two erstwhile belligerents.

Aerial Surveillance

Under the Sinai I Agreement, the United States carried out-high altitude aerial reconnaissance once every seven to ten days, or when specifically requested by Egypt, Israel, or UNEF II. These flights included U-2 and SR-71 aircraft, and while little has been released regarding the character of the imagery taken, the U.S. made the data available to both Egypt and Israel after analysis by American experts. These flights were made over both the buffer and arms limitation zones.

These U.S. reconnaissance flights and procedures were continued under the Sinai II Agreement, and were conducted over all of the areas relevant to the Agreement. Both Israel and Egypt were also permitted to conduct aerial surveillance activities as a means of national verification. They were allowed seven flights each week, and could fly freely to the buffer zone. Flights within the buffer zone were, however, limited to an area up to the middle of the zone and to an altitude not less than 4,570 metres. No public documentation exists as to the sensors used during the Egyptian and Israeli flights, although no restrictions were placed on sensors in the Agreement. With the signing of the Peace Treaty and the completion of the Israeli withdrawal, Egyptian and Israeli aerial reconnaissance flights were limited to zones A and D respectively — no evidence exists as to whether this imagery was ever shared. In this regard, the Parties were encouraged to utilise commonly available information for the purpose of exploring potential Peace Treaty deviations, even as they were permitted to collect their own data. Misunderstandings would be explored in a forum designed to mitigate against escalating charges of non-compliance.

UNEF II also made use of aerial reconnaissance, with aerial overflights monitoring activities in areas not easily accessible to ground-based troops. The force was equipped with four helicopters and three light aircraft — provided by Canada — and its personnel numbered 4,200, plus 124 UN Truce Supervision Organisation (UNTSO) observers. It was principally tasked with acting as a neutral observer and guardian of the buffer zone, as well as with the monitoring of the limited-force zones through on-site inspections, checkpoints, and patrols. With the signing of the Peace Treaty, its responsibility changed to one of operating checkpoints and observation posts and conducting patrols along the border and lines B and C. The force was also to be responsible for challenge inspections and bimonthly on-site inspections within zones A to D. With the 1979 refusal of the UN to extend UNEF II's mandate, these responsibilities fell to the SFM.

With the end of the UN mandate and the final Israeli withdrawal from the Sinai, the SFM acquired on-site inspection and low-altitude surveillance responsibilities as part of its mandate. In this regard, its task shifted from one of surveying the Gidi and Mitla passes to one of monitoring the number of personnel and treaty-limited equipment (TLE) possessed by Egypt and Israel within the limited force zones. However, its personnel number did not expand to meet these new requirements — staff remained at approximately 200 American civilians — and, as a result, the SFM was hard pressed to maintain adequate surveillance of the zones.

Still, the SFM did not deviate from its responsibility to carry out bimonthly on-site inspections in zones A and B and at the four Israeli technical sites. Instead, it emphasised even more strongly the aerial component of the inspection programme as a means of quickly and efficiently covering the greatest amount of territory possible. The timeframe for inspections was quite rigorous, with two days for inspections in zone A, one day for zone B, and one day for the four technical sites.

Safety was an important concern for the SFM, and this issue was clearly addressed following an initial incident in which the safety of a helicopter crew was jeopardised. That incident, and the logistical difficulties in coordinating inspections of permitted military installations with which the SFM was initially faced, resulted in the SFM Director demanding clear agreed inspection procedures. Subsequently, the cooperation provided by both Egypt and Israel represented an essential aspect of the ability of the SFM to carry out its verification responsibilities over such a large territory.

Operationally, Egyptian and Israeli authorities were notified of proposed inspections by direct teletype one week in advance of a reconnaissance flight.

Aerial reconnaissance was conducted over a two day period, generally one week before a scheduled on-site inspection. Each zone was further subdivided into sectors for planning and accuracy purposes and according to their particular topographic features and military concentrations. Zone A held seven sectors, while zone B held three. Both aerial reconnaissance and inspection teams were briefed on the contents and specific issues of each sector. Reconnaissance teams would fly over 2,700 air miles during a mission, with the northern part of the two zones typically surveyed the first day and the southern part the next. Aircraft normally flew at an altitude of between 800 and 1,000 feet, but often descended much lower so as to afford teams a closer examination of military formations and installations. Teams used hand-held cameras and other viewing devices. No landings were permitted during such reconnaissance missions, except at designated refuelling points. Where teams observed unusual objects or activities, the accompanying Egyptian or Israeli liaison officer was notified of the team's observation and a note would be made so that future ground inspections could clarify its nature. While the SFM sought to ensure that the same personnel were involved in the same reconnaissance and on-site inspection cycle, assignments were also rotated so as to mitigate against complacency and inattentiveness.[1]

The synergistic integration of aerial surveillance data into ground inspection activities contributed significantly to the effectiveness of the SFM's verification efforts. However, it must be remembered that the SFM's contribution was never meant to do more than supplement the early warning systems of Egypt, Israel, and UNEF II. Both Egypt and Israel often used the SFM as a neutral vehicle for the exploration of potential deviations suggested by their own or American national technical means (NTM). This ability to explore national compliance concerns within the framework of the Agreements and the Peace Treaty both strengthened the peace process and suggested that NTM could interact positively with a larger agreement. However, no public documentation indicates whether the SFM used U-2 or SR-71 aerial imaging, or whether this was going on simultaneously alongside SFM aerial reconnaissance. The success of the SFM's verification efforts lay to a large degree in the simplicity of its technology and the trust which both Israel and Egypt placed in its members. The SFM experience also demonstrated that aerial surveillance could only supplement, and not replace, information gathered on the ground. The use of NTM suggested that data from national flights could be positively integrated into an arms control framework, so long as agreements properly delineated the terms of the collection for such information.

The MFO has carried on the SFM's practice of utilising aerial surveillance and the transport of ground inspection teams by helicopter. In this regard, its mission timelines and sectoral divisions are generally a carryover from those of the SFM. Aside from watch stations manned by the three infantry battalions within the MFO, the MFO's Civilian Observer Unit (COU) — made up of fifteen U.S. nationals — engages in regular observation and verification missions through all

1. United States, Sinai Support Mission, *Watch in the Sinai*, 9–10.

four zones. Verification and surveillance activities are conducted not less than twice a month, and use both MFO vehicles and helicopters. Each inspection cycle also involves prior reconnaissance flights. The COU must also be ready to carry out additional verification activities within 48 hours of a request from Israel or Egypt. Both Egypt and Israel use their NTM as a guide to the need for further ground inspections, and the MFO also utilises American NTM in the design of its inspection plans. One special inspection has been prompted by an aerial reconnaissance flight. The first aerial support for the COU was provided by Australia and New Zealand — after their withdrawal in 1986, Canada provided nine CH-135 helicopters until April of 1990. At present, aerial support for the COU is provided by a French fixed wing aviation unit and an American rotary wing aviation unit. Canada currently provides an air traffic control unit to the MFO and the COU.

Unlike the SFM surveillance flights, MFO aerial surveillance teams may not make use of hand-held sensor equipment. However, because of the acceptance by Israeli and Egyptian forces of the principle of aerial surveillance, this has not provided the MFO surveillance teams with serious difficulties. MFO reconnaissance and verification flights are designed to interact synergistically with both roving ground patrols and ground inspections, although little open information exists as to the process for this interaction.

Surveillance and verification flights are required to carry Egyptian and Israeli liaison officers as observers and, despite the trust exhibited by these two parties toward the MFO, confidentiality still remains a sensitive issue. The presence of these liaison officers does inhibit communication amongst the MFO observers. The verification and reconnaissance activities of the MFO and the SFM focus upon specified locations and subjects, something which makes visual sweeps of an area difficult. Israeli and Egyptian authorities possess ultimate control over the flight corridors MFO aircraft may take, so that exhaustive area searches are precluded from the inspections activities. Nevertheless, all parties concerned continue to feel that the efforts of the MFO represent a substantive element in the verification of the Peace Treaty.

Much of the efficiency of the MFO — including its aerial surveillance activities — stems from its emphasis upon the importance of training programmes tailored to its particular mandate. Through the work of the Senior Staff Officer/Training and the Training and Advisory Teams, the MFO has established a training programme which establishes standardised inspection skills among its staff and ensures that, in the face of personnel turnover, the experiences of past knowledge are not lost to future inspection and other teams.

The start-up costs for the Sinai Support Mission stood at around $21 million (US), with approximately $84 million (US) spent in its operation between 1976 and the end of 1981. The cost of the MFO stood at $51,830,000 (US) in 1994 and $53,323,000 (US) in 1993, a considerable reduction from the $74,000,000 (US) spent in 1988. While no figures are given for the cost of aerial surveillance, in

1995, the MFO spent $1,411,000 (US) on transportation and $2,773,000 (US) on petroleum, oil, and lubricants. It should be mentioned that the United States, France, and Italy furnish the MFO with capital equipment at no cost, and share the maintenance and support costs of this equipment with the MFO. Furthermore, instead of paying fixed salaries for troops serving with the MFO, the organisation receives a credit for the costs seconded troops would have incurred at their home bases. Stated expenses are considered to be the net of expenses for equipment and supplies applied against these credits. Much of the cost of the MFO is shared equally between Egypt, Israel, and the United States. The financial burden shouldered by these three funds-contributing states — $16,349,000 (US) from each state in 1995 — has, however, been lessened somewhat by contributions from Japan, Germany, and other states.

Like the SFM, the ability of the MFO to integrate aerial reconnaissance and verification support into ground inspection activities has significantly strengthened its ability to effectively carry out its mandate. Like the SFM, the MFO relies for much of its success upon the political good will generated by the U.S.-supported peace process. With regard to the use of technology, its success is double-edged: while much of the willingness of the Israeli and Egyptian authorities to permit aerial overflights stems from the MFO's lack of surveillance equipment and, thus, its inability to engage in espionage, the MFO's success to date demonstrates what can be done with minimal sensor support in a supportive political environment. The willingness on the part of Egyptian and Israeli authorities to explore potential Peace Treaty discrepancies detected by either their own or American NTM within the MFO verification and consultative framework is important. It suggests that NTM and multilateral aerial reconnaissance can interact positively where the political will and a proper treaty framework exist.

The United Nations Special Commission on Iraq

Created in May 1991, in the wake of the post-Gulf War revelations of Iraqi research into and production of weapons of mass destruction, UNSCOM was designed to implement the terms of Security Council Resolution 687. This resolution called for, among other things, the identification, location, and destruction of Iraq's weapons of mass destruction and ballistic missiles with a range exceeding 150 kilometres, as well as the creation of an ongoing programme of monitoring compliance with these provisions. UNSCOM's activities were further strengthened by Security Council Resolutions 707 and 715, adopted in August 1991 and October 1991 respectively. These two resolutions were intended to respond to Iraqi obstructionism regarding inspections of ballistic missile capabilities and NBC weapons by demanding full disclosure and unconditional UNSCOM and IAEA access for inspections.

UNSCOM represents an unusual form of arms control agreement in that its mandate was imposed upon Iraq after the UN exhibited grave concerns regarding its proliferation activities. There was little desire on the part of the Iraqi government to comply with the terms of SCR 687 — indeed, the UN Security Council has faced an ongoing battle in attempting to strengthen verification measures in the face of selective Iraqi obstructionism vis-à-vis the missions of the IAEA and UNSCOM. In this regard, the UN has been able to draw upon the political support of the United States and other nations in imposing these arms control and destruction measures. There can be little doubt that without this political will, UNSCOM and the IAEA would be unable to continue with their verification, destruction, and future compliance mandates. Nevertheless, there has been some cooperation between UNSCOM and Iraqi officials on a day-to-day basis.

Aerial Surveillance

Although SCR 687 gave UNSCOM the right to carry out broad verification efforts in Iraq, at first little consideration was given to the potential of aerial surveillance. This changed with the first series of Iraqi efforts to interfere with inspection activities. Aerial documentation of the location of the calutrons that were removed from the Falujah Military Transport Command site after an Iraqi confrontation with an UNSCOM surprise inspection team suggested the value of such information. Satellite photography proved to be of limited utility, as such images could not always be procured when needed and could not be independently analysed by UNSCOM. Furthermore, the confidentiality of such information limited their distribution to the Office of the Executive Chairman and the chief inspectors.[2]

UNSCOM utilises a variety of forms of aerial surveillance. A U-2 aircraft provided by the U.S. provides high-altitude photography, while three CH-53 Sikorsky helicopters provided by Germany from UNSCOM's inception to mid-1996 assisted in low-altitude surveys, the scanning of potentially unsafe ground, and the quick transportation of ground inspection teams. Chile has subsequently provided five UH-1H helicopters. UNSCOM surveillance involves the verification of declarations, the inspection of declared or suspicious sites, and the monitoring of activity clearly or possibly related to Iraqi evasion of its responsibilities under SCR 687. It should be noted that while a number of aerial surveillance missions continue to involve the search for physical evidence of such proscribed equipment as ballistic missiles, the broader ongoing monitoring and verification effort of UNSCOM and the IAEA is becoming increasingly driven by assessments made from documentation gathered through exhaustive 'paper trail' searches, as well as other sources of information. While the subjects of such searches — biological and chemical weapons facilities, missiles, and nuclear processing infrastructures — may not be easily distinguishable by aerial surveillance, such monitoring is often a crucial element in evaluating the potential for non-compliant behaviour in such Iraqi tactics as announcing 'unilateral destructions' of relevant materials or in forbidding UNSCOM and IAEA teams access to 'sensitive' sites for certain periods of time.

---

2. Johan Molander, "Mandated Aerial Inspections: The Iraqi Case," Michael Krepon and Amy E. Smithson, eds., *Open Skies, Arms Control and Cooperative Security* (New York: St. Martin's Press, 1992), 149–150.

Security Council Resolution 707 explicitly states that the UN is permitted to conduct fixed-wing and helicopter flights throughout Iraq for the purposes of inspection, surveillance, aerial surveys, transportation, and logistics, and that these flights should not be interfered with in any manner. This resolution has been used by UNSCOM to support high-altitude reconnaissance in Iraq using the American U-2 aircraft. U-2 flights generally are conducted according to a regular schedule — at first once a week, increasing at times to three times a week. The Iraqi government is notified 72 hours in advance of the beginning of the 48-hour time frame for the flight. UNSCOM typically informs the U.S. of any areas it wishes imaged. Although the Executive Chairman is free to share these photographs with the Security Council, wider distribution may occur only after consultation with and the permission of the U.S. government. These flights were begun in September of 1991 and continue to the time of this writing. U-2 imagery was and still is used to monitor activities at declared sites as well as to search for sites or activities which might suggest the presence of further undeclared items, and utilises both sweeping and point imagery.

The Iraqi government has made a number of attempts to interfere with the U-2 overflights. In April of 1992 the government of Iraq demanded the cessation of all U-2 flights with the warning that further flights would place the aircraft and its pilot in danger. Although Iraq retreated from this position after a vigourous reminder from UNSCOM of its right to conduct such flights, the Iraqi government continues to refer to the U-2 as a 'spy plane' and protests each overflight as an incursion into its airspace.

Although useful in providing an initial overview of the territory to be inspected, high-altitude aerial surveillance suffers from some shortcomings. Most important, it has proven inadequate in discovering the myriad of underground and covered structures built by the Iraqis. Second, given the incomplete nature of the knowledge possessed by the West regarding Iraqi research into weapons of mass destruction, the limitations of early imagery interpretation may have contributed to the exaggeration of Western fears regarding aspects of the Iraqi research programme.[3] Still, high-altitude photography remains an important aspect of UNSCOM's effort to build a database of Iraqi sites which would assist UNSCOM staff and inspectors in assessing data, searching for potential locations of undeclared items, and planning future inspections. Furthermore, when combined synergistically with information from helicopter surveillance activities, it offers UNSCOM an important and unique means by which to identify and examine activities which could point to potential Iraqi non-compliance with the terms of SCR 687.

UNSCOM first introduced the use of helicopters in October of 1991 as part of a ballistic missile inspection. Much of the reason behind their use lay in UNSCOM's need for an information source which could increase the analytical

3. *Ibid*, 155.

utility of high-altitude aerial data in the face of inadequate Iraqi declarations. Aerial imagery from helicopters was superior in some ways to that provided by the U-2 as it offered greater opportunities for oblique photography as well as the possibility of examining a site from a variety of angles.

In the summer of 1992, Iraq sought to hinder helicopter surveys by demanding that they not overfly Baghdad or other sensitive sites, that they follow flight patterns defined by Iraq, and that they not engage in area surveys. These tensions increased in 1993 with the Iraqi threat to shoot down any UNSCOM helicopter which was engaged in aerial surveillance in support of a ballistic missile inspection team. That issue was only overcome with the assistance of Security Council and member political and military pressure. During a series of on-site inspection activities in 1996, Iraq sought to impose restrictions upon aerial surveillance activities carried out in support of ground inspections. In June of that year, Baghdad went so far as to train anti-aircraft guns on UN helicopters surveying sites which had been declared off-limits to UNSCOM inspectors. Still, Iraqi efforts at interference can, perhaps, be realistically seen as less of an attempt to prevent helicopter surveillance than an effort — to date unsuccessful — to establish agreed-upon parameters which would narrow UNSCOM's flexibility in this regard.

These helicopter flights continue to assist inspection teams by allowing them to rapidly cover great distances to reach otherwise remote sites, search for suspected sites, and survey the areas surrounding declared sites. Another part of their utility stems from their ability to serve as a platform for a variety of sensors. Despite earlier protestations, Iraq chose, in September 1993, to accept the deployment of gamma-radiation sensors on board these helicopters, which improved UNSCOM's ability to explore for unusual concentrations of gamma radiation in a speedy and cost-effective manner. Any findings can be followed up by ground inspections, demonstrating the important synergy of these overflights with on-site inspections. Iraq has also accepted UNSCOM's intention to use ground-penetrating radar and forward-looking infrared radar, despite previous concerns regarding the use of high-technology surveillance equipment. Aside from the improved vantage point that helicopters offer inspection teams,[4] these technologies have enabled UNSCOM to compensate for the Iraqi governments incomplete declarations with improved surveillance capabilities. The ability of helicopters to provide a highly mobile platform for low-level aerial surveillance using a variety of sensor types also strengthens UNSCOM's ability to successfully interpret and act upon the clues provided by U-2 imaging data, particularly where the issues involved are of a time-sensitive nature. The importance of this aerial support to the broader UNSCOM ongoing monitoring and verification (OMV) effort was clearly demonstrated when the cutback of one fixed-wing and one helicopter aircraft forced the Commission to engage in a number of difficult mission

---

4. In one case, helicopter-borne inspectors on an inspection of chemical weapons storage bunkers accidentally discovered three undeclared Scud missiles near the inspected site.

trade-offs. The Commission made it clear at the time that it considered the aerial support given by the German government to constitute the minimum number of aircraft with which it could carry out its mandate.

While the ability to field aerial surveillance platforms and technologies was an important issue for UNSCOM, the capacity to interpret data gleaned from these sources was no less salient. The evolution of the Information Assessment Unit (IAU) located within the headquarters staff in New York represented a critical step in this regard. Created in February 1992, it was designed to enable UNSCOM to prepare for the responsibility of monitoring Iraqi future compliance with SRC 687. Previously, UNSCOM had relied upon information provided voluntarily by Member States to corroborate and interpret Iraqi declarations — information which, in the light of ongoing revelations regarding the Iraqi WMD research programme, proved to be not wholly accurate. Furthermore, operations and planning functions within UNSCOM's main office were often intertwined, and there was a distinct tendency to divide the organisation into working groups headed by a chief inspector for the purpose of developing and launching each specific inspection effort. While this ad hoc organisational format certainly proved effective in creating strong inspection efforts, its lack of structure mitigated against an institutional memory or follow-up activities, as well as analytical research.[5] While the IAU initially operated as a repository for inspection reports and U-2 and helicopter imagery, as well as a collection agency for information provided by contributing governments, it grew into a body capable of interpreting collected information for the purpose of planning future inspections in response to the demands facing UNSCOM. This analytical capability proved essential for UNSCOM in maintaining its direction in the face of both continuous Iraqi assertions that all WMD research and activities had been declared or discovered, and in addressing the position of certain states that UNSCOM had discovered only a fraction of the Iraqi research programme.

The establishment of the Baghdad Monitoring and Verification Centre (BMVC) and the Baghdad-based Aerial Inspection Team represented other steps in the effort to formalise the gathering and management of information gleaned from, among other activities, aerial surveillance and support. The BMVC, established in 1994, serves as a local focal point for, among other things, communications with the UNSCOM and IAEA headquarters in New York and Vienna, the real-time monitoring of various sites through remote-controlled cameras, medical services, biological and chemical laboratories, and the offices for the aerial inspection team and the biological, chemical, missile, export/import, and nuclear monitoring groups. The Aerial Inspection Team was created to give a greater degree of consistency and continuity to UNSCOM's aerial surveillance efforts,

---

5. Johan Molander, "The United Nations and the Elimination of Iraq's Weapons of Mass Destruction: The Implementation of a Cease-Fire Condition," Fred Tanner, ed., *From Versailles to Baghdad: Post-War Armament Control of Defeated States* (Geneva: United Nations Institute for Disarmament Research, 1992), 144–145.

and by 1996 had taken part in more than 600 missions — in contrast to the approximately 270 missions flown by the U-2. Experts from the other monitoring groups — now with their own permanent team structures — accompany the team where necessary as a means of further focussing the aerial surveillance effort and ensuring the smooth exchange of information and concerns between the teams. The BMVC further supports the work of the Aerial Inspection Team through the provision of a photographic development laboratory at the Centre, something which allows the team to quickly process and assess photographic data in the support of other group activities. UNSCOM has also participated in the development of a remote monitoring/information management workstation with Pacific Northwest as a means of providing inspection teams with multimedia information capabilities across such diverse data types as maps, aerial imagery, still and motion photographs/videos, texts, and tabular data on a location-referenced basis. It is expected that this technology will further augment the decision-making capabilities of inspection teams and reduce reaction times under what continues to be a highly time-sensitive mandate.

The United States and Germany have absorbed the costs of maintaining the U-2 aircraft and CH-153 helicopters respectively. With the withdrawal of German air support in 1996, the Chilean government presently bears the financial burden of its rotary aeroplane contribution. These sums are not insignificant; one analysis suggested that the bulk of the money spent by the U.S. in Iraq in fiscal year 1993 — roughly $19 million — went to the maintenance of the U-2 and the cost of developing film,[6] while the German government has been thought to have spent approximately $50 million (US) on aerial support for UNSCOM over the last four years. In the context of a yearly UNSCOM budget of $24–28 million (US), these sums demonstrate a considerable national support of UNSCOM's surveillance and inspection activities. It should be noted that UNSCOM is a unique entity, using national specialists at no cost and, in return, covers the incidental expenses of these individuals. Furthermore, it has generally been suggested that the annual cost of operating the IAU with adequate equipment stands at roughly $2 million (US) — a sum which, in comparison with other verification activities, remains quite small considering its impact upon the effectiveness of UNSCOM's inspection efforts.

It must be emphasised that UNSCOM's success to date in overcoming Iraqi evasion and resistance to their verification efforts cannot be completely placed at the feet of this organisation's ability to utilise aerial platforms and sensors in the preparation, direction, and support of inspection efforts. Much of UNSCOM's success can be attributed to the willingness of particular UN Member States to use their political and military influence to compel Iraq to accept increasingly intrusive sensors and inspection activities. An embargo of Iraqi oil also has done much to increase the willingness of the 'cash-strapped' Iraqi government to

6. Andrew Weinschenk, "U.S. Still Foots the Bill for Smoking Out Iraqi Weapons," *Defence Week*, 12 October 1993, 3.

tolerate UNSCOM surveillance and inspection support activities. Iraqi intransigence during the summer of 1993 was overcome, in large part, by the threat of a continuation of the oil embargo, as well as by further military operations. Clearly, UNSCOM continues to carry out its verification activities in Iraq within a larger cocoon of Western 'carrots' and 'sticks' — indeed, without such support, UNSCOM would likely have enjoyed little of its present-day success.

Conclusion

As mentioned above, any comparative assessment of the aerial surveillance activities of the SFM/MFO and UNSCOM must take into consideration some important differences in their missions and larger political contexts. While the MFO operates in a political environment which is broadly supportive of its work, the inspection teams of UNSCOM typically find themselves harassed by an Iraqi government generally hostile to the mandate of this organisation. Furthermore, while the missions of the SFM and the MFO generally involved the monitoring of various bases to ensure that Peace Treaty limits on personnel and equipment were respected, UNSCOM is engaged in a more difficult mission of attempting to ascertain the level of Iraqi research into and holdings of equipment, weapons, and components proscribed by the terms of SCR 687 — typically through searches of seized documentation and subsequent efforts to evaluate the accuracy of previous assessments of the Iraqi material balance.

Despite these differences, it is possible to draw some parallels between these two verification experiences regarding the use and utility of aerial surveillance. First, while high-altitude surveillance can serve as an important element in aerial verification, even in those arenas characterised by a politically stable strategic structure, such activities should be complemented by low-altitude surveillance/monitoring efforts. While high-altitude surveillance provides an overview of the layout of a particular region, low-altitude surveillance and monitoring activities provide a clearer understanding of the activities occurring at sites of interest within that region. Both the MFO and UNSCOM experiences suggest that low-altitude surveillance and monitoring activities are the most effective means by which to investigate time-sensitive phenomena, as well as to explore suspicious anomalies. In the execution of such aerial surveillance, rotary-wing platforms seem to be the most flexible of aerial platforms, with their ability to take off and land in difficult terrain, transport small inspection teams quickly, and cover relatively large distances to reach sites which are difficult to reach otherwise. With regard to the use of national technical means on the part of the SFM, the MFO, and UNSCOM, it is clear that NTM and multilateral aerial reconnaissance for verification can interact positively where the political will and a proper treaty framework — or some other, more informal, but negotiated procedure — exist.

Second, where high-altitude aerial data can be combined with low-altitude aerial examinations of areas of particular interest, the need to cover large geographic areas need not be a hindrance to adequate surveillance and monitoring efforts. Indeed, aerial surveillance and monitoring teams may often find their effectiveness more constrained by political sensitivities than technological limitations.

Third, the costs of supporting fixed- and rotary-wing aircraft were somewhat greater for UNSCOM, relative to the context of its budget, than the costs incurred by the SFM and the MFO. However, given the critical role aerial surveillance and support has played in the execution of UNSCOM's mandate, it would appear that the more challenging the verification mission, the more the 'return on investment' for aerial surveillance activities. This would seem to suggest that additional financial investments are more than recouped through improved access to information.

Fourth, the attention given by the MFO and UNSCOM in training and standardising inspection procedures speaks to the importance of such activities in ensuring that a broader continuity exists among the inspection teams — one that can bridge the confusion created by transfers in personnel and the inevitable difficulties new staff have in quickly adapting to different tasks. Standardised training procedures and established team structures generate a cohesiveness critical to the planning of missions which seek to build upon past experiences, adapt to various lessons, and anticipate future challenges. In other words, it establishes a more stable and adaptable institutional memory.

Finally, while the financial uncertainty inherent in relying upon national contributions for aerial platforms has been a cause for concern within the SFM, the MFO, and UNSCOM, the presence of these national contributions may also bring with them some advantages by affording aerial inspection teams a measure of political protection or respect which an international organisation alone might not offer. The American aerial presence in the Sinai has served as a tangible reminder of the United States' broader commitment to the peace between Egypt and Israel, and the various other national contributions to the MFO have underscored a multilateral willingness to support an agreement that brings a broader stability to the Sinai region. While the Iraqi attitude toward foreign nationals flying UNSCOM aircraft may lack some of this respect or common interest, the presence of these nationals supporting the Commission's activities adds an additional element of political uncertainty with every threat the Iraqi government issues against aerial inspection teams. While it would be extremely difficult to gauge the seriousness with which the Iraqi government considers the dangers of potential retribution brought on by targeting foreign nationals engaged in supporting UNSCOM missions, past experience has demonstrated that the Commission enjoys a considerable level of political support amongst the erstwhile members of the Gulf War coalition — a support that has, at times, been used to extract compliance on the part of the Iraqi government with the terms of SCR 687.

There are, however, some differences between the experiences of the SFM, the MFO, and UNSCOM which must also be considered. While the process of standardisation and institutionalisation within the SFM and MFO generally represented an effort to ensure consistency of inspection skills across teams and over time to new units, UNSCOM has gone further to explore means by which to ensure the efficient collation and analysis of data for the purposes of influencing future decisions regarding inspection activities. This additional step — one that goes beyond the attempt to create an institutional memory — stems in part from the more data-oriented nature of compliance assessment activities within the context of the Commission's mandate, and the need to adapt inspection missions to a constantly shifting compliance environment in as timely a manner as possible. UNSCOM staff have, out of necessity, built an elaborate, highly-structured and synergistic system of data management not needed within the SFM and the MFO. It should be recognised that analytical independence is important in environments of political or strategic confrontation, where an international verification organisation may find itself caught between the inspected government's claim of openness and foreign governmental assertions of non-compliance. However, future verification efforts similar to UNSCOM's mandate — the search for inconsistencies in data and potential acts of non-compliance in sites as yet unknown — will also need to emphasize the creation of organisations which support the effective management and assimilation of complex data sets. In this regard, they should look to the structures established by UNSCOM for the efficient exchange of information and the improvement of timely decision-making procedures for guidance.

Chapter Thirteen

Spaceborne Imagery: A Universal, Effective, and Cost-Efficient Tool for Ongoing Monitoring and Verification

Phillip J. Baines*

Introduction

Outer space is a region that offers effective and cost-efficient solutions for the ongoing monitoring and verification (OMV) measures required of non-proliferation, arms control, and disarmament (NACD) agreements. The stabilising effects of military spaceborne reconnaissance on the conduct of the affairs of nations are a matter of historical record. The universal commercialisation of high-resolution, spaceborne imagery offers innovative prospects for the NACD verification mission. Satellite imagery may see increased utilisation for OMV as cost-conscious governments seek means for increasingly effective verification at ever lower levels of expenditure. The increased utilisation of Earth observation assets for NACD verification may well require legally-binding instruments to govern their use and to protect them from deliberate harm.

Historical Developments

An early mission of the space age was the collection of spaceborne imagery for assessing the military balance between the two superpowers. American Corona and the Soviet Kosmos series of satellites continuously surveyed the globe, detecting and classifying targets of strategic interest. The resolution afforded by the first successful Corona mission in 1960 was 8 metres and, by the time of the last Corona mission in 1972, the resolution was improved to 2 metres. The development of these systems allowed the United States and the Soviet Union to secure bilateral arms control agreements such as the Strategic Arms Limitation Talks (SALT) I and the Anti-Ballistic Missile (ABM) Treaty in 1972 during the adversarial days of the Cold War. Follow-on missions for these classified systems not only included verification of arms control agreements, but also the collection

---

* This chapter is an edited version of a presentation made to the Fourteenth Annual Ottawa NACD Verification Symposium, "Cyberspace and Outer Space: Transitional Challenges for Multilateral Verification in the 21st Century," 12–15 March 1997, Montebello, Quebec, Canada. The opinions expressed in this paper are those of the author and do not necessarily reflect those of the Department of Foreign Affairs and International Trade or of the Government of Canada.

of information suitable for military analysis of new weapons developments. These military systems were also designed to provide targeting information and damage assessment information during times of conflict. Billions of dollars were spent by both of the superpowers to acquire and operate these space systems.

The ABM Treaty of 1972 set the fundamental principles of arms control verification using spaceborne assets under the umbrella of national technical means (NTMs). These principles included: the use of national technical means of verification in a manner consistent with generally recognised principles of international law; prohibiting interference with the national technical means of verification; and, discouraging deliberate concealment measures which impede verification by national technical means. A common understanding arising from the SALT II Treaty further added the verification principle of functionally related observable differences (FRODs). FRODs are differences in the observable features of delivery platforms that show their intended function. Thus, deliberate distinguishing features were incorporated on strategic aircraft as a cooperative measure for enhancing the effectiveness of verification by national technical means.

Once the underlying space technology had been developed in the military domain, the natural diffusion of this knowledge saw application in the commercial domain. SPOT 1, the world's first commercial remote-sensing satellite, changed the relationship between the world's peoples and the intelligence agencies of the United States and the Soviet Union when it was lofted into orbit by an Ariane rocket on 22 February 1986. Owned by French, Swedish, and Belgian interests, SPOT 1 placed at the disposal of the public a 'virtual' reconnaissance satellite that could recognise objects as small as 10 metres across. No longer was a national 'spy' satellite required to detect and classify military installations. Several organisations exploited this readily available, commercial information resource. The Soviet Chernobyl fire, the Soviet missile early warning radar station at Krasnoyarsk, and the Libyan chemical weapons facilities at Rabta are events that rest securely in the public domain, thanks in part, to commercial remote-sensing satellites.

Commercial remote-sensing satellites developed doggedly over the past decade as each satellite could cost up to (US) $500 million and a further (US) $100 million to launch. The missions for these unclassified systems were centred on the provision of data suitable for Earth resource monitoring, but not necessarily for arms control verification. Despite the moderate resolution of these civilian systems, countries with their own national technical means continued to exploit these unclassified systems. A prime example is the U.S. Department of Defense's *Soviet Military Power* publications of the 1980s, in which the conclusions of classified image analysis reports may have been revealed through discussion of SPOT imagery. In addition, the Gulf War experience saw commercial imagery provided to commanders in the field faster than from NTMs because of the unclassified nature of the data.

Commercialisation of space continued with vigour at the end of the Cold War when in 1991, Russia began selling satellite imagery with 2 metres resolution obtained by degrading the imagery from its own national technical means. India also launched its indigenous IRS-1C satellite with a resolution of 5.8 metres. In response to these competitive forces, the United States authorised commercial remote-sensing satellites with a 1 metre resolution capability in 1994. These satellites are expected to become operational as early as December 1997. **Table 1** shows the universality of commercial remote-sensing satellites before the close of the twentieth century. With such spaceborne imagery offered on a strictly commercial basis, innovative ideas may soon enter service for NACD OMV.

The verification principles enshrined in bilateral treaties between the United States and now the Russian Federation, however, are not so clearly codified in multilateral treaties. The Outer Space Treaty of 1967 may establish the principle of freedom of use as it pertains to remote-sensing conducted from space, but the Treaty contains no direct references to any kind of observation from space. Since the U.S. and the former USSR have conducted spaceborne reconnaissance without significant protest from other countries in the world, it is argued that these activities may now be legitimised through international customary law. With the capability of the commercial remote-sensing systems restricted to militarily questionable values in the past, and with the secrecy surrounding the use of NTMs in essentially a bilateral context, this ambiguous legal regime may have been sufficient. As the next generation of high-resolution, commercial spaceborne imaging becomes available, however, legally-binding rules may need to be established to govern the use of the imagery and to protect the spaceborne sources of the imagery from deliberate harm.

Applications

Application of high-resolution, commercial spaceborne imagery may be postulated for four major multilateral non-proliferation, arms control, and disarmament agreements: the Chemical Weapons Convention (CWC), the Bacteriological and Toxin Weapons Convention (BTWC), the Nuclear Non-Proliferation Treaty (NPT), and the Comprehensive Test Ban Treaty (CTBT).

The CWC is a Treaty that requires rigorous on-site inspection provisions for effective verification. Chemical weapons facilities may possess characteristic signatures in overhead imagery that could trigger the request for a challenge inspection. Dual-use of facilities may, however, complicate verification by spaceborne means alone. As such, the CWC does not explicitly endorse the use of overhead imagery for verification purposes. Signatories to the CWC could, however, exploit commercial satellite imagery for locating chemical facilities, and for the requisite preparation of site plans for these facilities, before the conduct of an on-site inspection. Satellite imagery may, therefore, play a supporting role in the verification of the CWC.

Table 1: Universal Remote-Sensing Satellites Circa 2000

REGION	COMPANY	SATELLITE / SENSOR	RESOLUTION	LIFE	YEAR
EUROPE	SOVINFORMSPUTNIK	KOSMOS KVR-1000	2 M	45 DAYS	1991,...
	GLAVKOSMOS	ALMAZ-1,-2,-3 SAR-10	15 M X 15 M	2 YRS	1991,'98,'01
	SPOT IMAGE CORP.	SPOT-1 TO SPOT-4 HRV	10 M	3 & 5 YRS	1986,'90,'93,'98
		SPOT-5,-6 HRV	5 M	5 YRS	2001,'04
	ESA	ERS-1,-2 SAR	30 M X 30 M	3 YRS	1991,'95
		ENVISAT ASAR	30 M X 30 M	5 YRS	1999
ASIA	NASDA	ADEOS AVNIR	8 M	3 YRS	1996
		JERS-1 SAR	18 M X 18 M	2 YRS	1992
		ALOS PRISM	2.5 M	?	2002
		ALOS AVNIR-2	10 M	?	2002
		ALOS PALSAR	10 M X 10 M	?	2002
	ISRO	IRS-1C,-1D PAN	5.8 M	3 YRS	1995,'97
AMERICAS	SPACE IMAGING CO.	SIS-1,-2 PAN	1.0 M	5 YRS	1997,'98
	ORBIMAGE CO.	ORBVIEW-3 PAN	1.0 M	5 YRS	1999
	EARTHWATCH INC.	QUICKBIRD-1,-2 PAN	0.82 M	5 YRS	1998,?
	RADARSAT INTERNATIONAL INC.	RADARSAT-1,-2 SAR	8 M X 8 M	5 YRS	1995,'00

Source: Compiled from the World Wide Web as at 1 April 1997.

The BTWC is the first Treaty to ban a complete class of weapons of mass destruction. It was concluded without rigorous verification provisions. A review conference of the BTWC will seek to redress this issue given the successful conclusion of the Chemical Weapons Convention. As bacteriological weapons development can be conducted within the confinement of a single, nondescript building, the use of overhead imagery may not be particularly suited for direct verification of the BTWC. However, determining the location of isolated, secure facilities, like aerosol dispersion test ranges, and the further preparation of site plans for these facilities, may be of particular use before the conduct of an on-site inspection. Overhead imagery, as in the CWC, may play a supporting role in the verification of the BTWC. The verification regime for the BTWC is not, however, expected to offer a radical departure from the CWC provisions.

The Iraq experience has drawn sharp criticism of the safeguards provisions in the NPT concerning undeclared facilities. The magnitude of the Iraqi nuclear weapons programme, as uncovered by the UN Special Commission (UNSCOM), was found to rival the scope of the Manhattan Project. And yet, Iraq was a signatory to the NPT and had placed its declared facilities under the safeguards provisions of the International Atomic Energy Agency (IAEA). The IAEA's "93+2" safeguards programme may redress this situation with "no notice" inspections of sites whose existence is derived from new reporting requirements. These safeguards may not, however, fully address the possibility of a parallel, clandestine fuel cycle. The review conference for the NPT may well require the adoption of additional verification methods, such as spaceborne imagery, to provide a means to monitor for undeclared activities. The IAEA has conducted preliminary studies illustrating the utility of spaceborne imagery for such purposes.

The CTBT requires a verification regime based on four monitoring methods: seismic, hydroacoustic, radionuclide, and infrasound, as well as the use of on-site inspections. The Treaty allows for overflights to be conducted during on-site inspections for the purposes of providing the inspection team with a general orientation of the inspection area, narrowing down, and optimising the locations for ground-based inspection, and facilitating the collection of factual evidence using specified equipment. The Treaty also obligates State Parties to examine the potential of additional monitoring technologies, such as satellite monitoring, to develop specific measures to enhance the efficient and cost-effective verification of the Treaty. The CTBT may offer the best prospect of incorporating spaceborne imagery as an integral part of a multilateral verification regime in the future.

With the prospect of high-resolution, commercial remote-sensing satellites offering a significant capability suitable for verification, the question arises as to how to harness these systems for NACD purposes. Several satellite imagery utilisation models may be postulated:

- national technical means (NTM);
- multinational technical means (MTM);

- treaty technical means (TTM);
- international technical means (ITM); and,
- non-government organisation monitoring (NGM).

The first model for the utilisation of high-resolution, commercial satellite imagery is for countries that do not possess their own national technical means to use such systems as a surrogate. In this scenario, states would buy the satellite imagery from service providers and analyse the data at national data centres for their own monitoring purposes. The compliance analysis produced by this means may then be used to develop policy on a strictly national basis, a bilateral basis, or a multilateral basis. In the latter case, multilateral treaty organisations may need to be prepared to accept such data supporting a challenging state's claim of non-compliance.

A second model of utilisation of high-resolution, commercial satellite imagery is for like-minded states that do not possess their own national technical means to use the capability collectively as a multinational technical means (MTMs). Examples of these cooperative approaches are the putative Franco-German *et al.* Helios and Horus classified satellite systems, and the Western European Union (WEU) satellite centre at Torrejon, Spain. Here, like-minded states set monitoring task priorities and jointly analyse the data for each country's independent political judgement. In this model, the costs will be shared by the participants and a stronger voice of condemnation of non-compliance may arise from a coordinated collection of states.

The third model is for a treaty organisation to procure the images and analyse them at a data centre owned by the organisation. This would be similar to the Canadian Paxsat proposals of 1987, except that the space segment is made unnecessary by existing commercial satellites. This approach legitimises the use of satellite imagery through explicit treaty provisions. The scenario also opens the compliance analysis process to all signatories for transparency and provides a controlled environment necessary to ensure the confidentiality of the derived analysis. Monitoring costs would also be shared by all states that are parties to the treaty. In addition, operating in an international environment overcomes knowledge or technology transfer limitations through training and collective ownership of the necessary facilities.

A fourth model is for a global agency to procure the images and analyse them at a data centre owned by the agency. This would be similar to the first two phases of the French proposal of 1978 for an International Satellite Monitoring Agency (ISMA). In this model, a centralised data centre performs the image analysis for a multitude of treaty organisations or states. As cost impacts are increasingly felt for expanding multilateral treaty organisations with similar OMV missions, a reformed UN may find a centralised verification agency even more effective, universal, and cost-efficient for the multilateral verification mission. A central verification agency, operating under the appropriate legal means, would avoid the pitfalls of the initial ISMA proposal. Furthermore, the operational experience gained by the WEU satellite centre and the reaffirmation of third party verification in the IAEA, may dispel the operational arguments arrayed against the initial ISMA proposal. The main advantage of this model is the increased cost-efficiency gained through the elimination of image acquisition and interpretation redundancy spent by individual states, collections of states, or a multiplicity of treaty organisations.

The last model does not foresee the use by governments, but instead, non-governmental organisations (NGOs) air the data. Examples here include the Swedish Media Centre's discovery of a Soviet laser weapons facility at Dushanbe. In this scenario, what would be the impact on the conduct of international diplomacy, or domestic politics, if the public makes better use of available information than many national governments? Governments may well need to examine the consequences of not making the best use of available information.

At one time or another, each of the above models was predicated on the use of commercial technology to build dedicated satellite systems for the NACD ongoing monitoring and verification mission. Today, high-resolution, spaceborne imagery offered on a strictly commercial basis may negate the need for states, NACD organisations, or NGOs to procure the space segment portions of the verification system. Instead, these organisations may concentrate solely on the reception, processing, and analysis of the spaceborne data. Whereas ten years ago a few billion dollars was needed to perform an NACD mission, now the cost of a mission may fall to within a few millions of dollars.

Effectiveness

The effectiveness of commercial satellite imagery for multilateral NACD verification will depend, *inter alia*, on the spatial resolution, the temporal resolution, and associated costs for these systems. **Table 2** illustrates the resolution requirements for photographic interpretation. At 1 metre resolution, vehicles, supply dumps, and artillery should be detectable in satellite imagery. Detection is the ability to discern that an object of possible interest is located in the image. Interpreters of this satellite imagery probably can classify missile, radio, and radar sites into general categories. Classification is the ability to discern, for example, whether an object is a fixed or rotary winged aircraft. Organised troop units, airfields, and aircraft should be recognisable at 1 metre resolution. Recognition is the ability to discern that an objects particular class can be definitely established. For example, the aircraft is definitely a fighter as opposed to a transport. Finally, at 1 metre resolution, interpreters probably can identify humanmade infrastructures like rail and port facilities. Identification is the ability to discern the specific type of object within its class. For example, the fighter aircraft is a CF18 versus a CF105. Sample airborne imagery provided by the Space Imaging Company confirms the detection capability for vehicles with a 1 metre resolution capability. Recall that this level of clarity is an order of magnitude improvement over Spot 1 and twice as good as the military systems of 25 years ago used to verify significant arms control agreements.

Table 2: Resolution Requirements for Photo Interpretation

TARGET	PHOTOGRAPHIC RESOLUTION (M)			
	DETECTION	CLASSIFICATION	RECOGNITION	IDENTIFICATION
VEHICLES	1.5	0.6	0.3	0.05
SUPPLY DUMPS	1.5	0.6	0.3	0.03
ROCKETS and ARTILLERY	0.9	0.6	0.15	0.05
SURFACE SHIPS	7.6	4.6	0.6	0.3
MISSILE SITES (SSM/SAM)	3	1.5	0.6	0.3
RADIO SITES	3	1.5	0.3	0.15
RADAR SITES	3	0.9	0.3	0.15
AIRFIELD FACILITIES	6	4.6	3	0.3
ROADS	9	6	1.8	0.6
TROOP UNITS	6	2	1.2	0.3
AIRCRAFT	4.6	1.5	0.9	0.15
COMMUNICATION HEADQUARTERs	3	1.5	0.9	0.15
PORTS and HARBOURS	30.5	15	6	3
RAILWAY YARDS	30.5	15	6	1.5
URBAN AREAS	30.5	4.6	3	1.5

Source: B. Jasani, "Arms Control and Conflict Observation Satellites," *Space Policy* 1:14 (November 1985). 363-368.

Table 3 shows the temporal coverage for a selected set of civilian remote-sensing satellites operating independently of one another. Satellites with synthetic aperture radar payloads can provide global accessibility every five days while satellites with electro-optic payloads can provide global accessibility every three days. Operated as a constellation of satellites, these missions should approach every-other-day accessibility for anywhere in the world, and complete imaging of the Earth's surface every season. The coverage offered by this plethora of civilian remote-sensing satellites may even approach the current U.S. operational reconnaissance constellation of two radar imaging satellites and three-to-five electro-optic satellites.

As the number of on-site inspections proliferates with each newly negotiated treaty, the total costs for these provisions will escalate. In a climate of fiscal responsibility, a resistance to these provisions may grow, and a more cost-efficient system may be sought to screen events before committing to a costly on-site inspection. **Table 4** illustrates indicative costs for on-site inspections, overflights, and satellite imagery drawn from a variety of sources normalised by the area of coverage. One study placed the cost of an on-site inspection at as much as (US) $60,000 (1990) for the challenging country and (US) $34,000 (1990) for the host country.[1] A typical base was assumed to occupy a region of 10 kilometre by 10 kilometre. Other studies of the Open Skies Treaty considered a fleet of eight Dash 8 Series 300 aircraft for the surveillance of Europe with synthetic aperture radar and infrared electro-optic imagery. This system was expected to cost approximately (US) $330 million (1989) over twenty years.[2] Satellite imagery costs for RADARSAT and SPOT data were provided by a Canadian commercial entity authorised for such sales. Interpretation costs were based on rough order of magnitude estimates provided for a functional analysis of a single image by a newly formed commercial entity specialising in military interpretation of satellite imagery. In addition, the WEU satellite centre experience illustrates that a multinational monitoring system may be operated for as little as (ECUS) $11 million (1996) per year.[3] Satellite imagery may well provide the most cost-efficient monitoring data for multilateral NACD agreements requiring a global reach.

Challenges

The widespread use of high-resolution, commercial satellite imagery for multilateral NACD verification faces several challenges. Foremost among these challenges is the fact that spaceborne imagery, whether military or civilian, is not

---

1. D. Lockwood, "How Much Will Verifying START Cost?," *Arms Control Today* 20:8 (October 1990), 18.
2. J. King, "Airborne Remote Sensing For Open Skies: The Platform," in M. Slack and H. Chestnutt, eds., *Open Skies: Technical, Organizational, Operational, Legal, and Political Aspects* (Toronto: Centre for International and Strategic Studies, York University, 1990), 21–34.
3. Anon, "WEU and Helios 2," Assembly of the Western European Union, Document 1525, 14 May 1996.

Table 3: Global Coverage Capabilities of Civilian Remote-Sensing Satellites

SATELLITE / SENSOR	INCLINATION (DEGREES)	ALTITUDE (KM)	FIELD OF REGARD (DEGREES)	SWATH WIDTH (KM)		GLOBAL COVERAGE (DAYS)	
				ACCESS	IMAGE	ACCESS	IMAGE
ALMAZ-1,-2,-3 SAR-10	72.7	380	L[26.60,55.43] R[26.60,55.43]	2 x 320	100	1 TO 4	?
ERS-1,2 SAR	98.5	781	R[23.00 CA]	100	100	11	35
JERS-1 SAR	97.7	577	R[35.00 CA]	74	74	43	44
RADARSAT-1,-2 SAR	98.6	797	R[20,49]	510	Fine 45 Std 100 Scan 510	1055	24 ? 5
ENVISAT ASAR	98.5	781	R[15,45]	400	Image 100 Scan 400	11 ?	35 ?
ALOS PALSAR	98.2	700	R[20,55]	600	Image 70 Scan 250	? ?	45 ?
COSMOS KVR-1000	67.1	220	[-22.2,22.2]	180	40	?	?
SPOT-1 to -6 HRV	98.7	822	[-27,27]	944	60	5	26
ADEOS AVNIR	98.6	797	[-40,40]	1572	80	3	41
ALOS PRISM	98.2	700	[-1.5,1.5]	54	35	45	45
ALOS AVNIR-2	98.2	700	[-40,40]	1370	70	3	45

SATELLITE / SENSOR	INCLINATION (DEGREES)	ALTITUDE (KM)	FIELD OF REGARD (DEGREES)	SWATH WIDTH (KM)		GLOBAL COVERAGE (DAYS)	
IRS-1C,-1D PAN	98.7	817	[-26,26]	904	70	5	24
SIS-1,-2 PAN	98.1	680	[-26,26]	690	11	4	247
ORBVIEW-3 PAN	97.3	470	[-45,45]	996	8	3	370
QUICKBIRD-1,-2 PAN	52	600	[-30,30]	720	22	4	148

Source: Compiled from the World Wide Web as at 1 April 1997.

Notes: Field of Regard is cross-track gimbal angle for EO and incidence angle for SAR payloads.

Table 4: Cost Comparison of On-site, Overflight, and Satellite Imagery OMV Methods

METHOD	AREA (KM^2)	ACQUIRE COST ($US 1990)	ACQUIRE COST/AREA ($US/$KM^2$)	INTERPRET COST/AREA ($US/$KM^2$)	TOTAL COST/AREA ($US/$KM^2$)
ON-SITE INSPECTION	BASE 10 X 10	94000	940	0	940
OVERFLIGHT IR	SORTIE 2 X 2,035	12600	3.1	1.15	4.25
OVERFLIGHT SAR	SORTIE 25 X 3,785	19950	0.21	1.15	1.36
SATELLITE IMAGERY EO	IMAGE 60 X 60	3085	0.86	1.15	2.01
SATELLITE IMAGERY SAR	IMAGE 50 X 50	3500	1.4	1.15	2.65

Notes: U.S. inflation for 1988 to 1997 was assumed at 2% per year. Exchange rate for Canadian dollars into U.S. dollars (1997) is 0.73060.

explicitly endorsed as an accepted means of verification within the multilateral treaty provisions.

Use of military reconnaissance satellites for multilateral verification has encountered several problems. Foremost is the problem that NTM systems tend to be the most secretive systems in a state's arsenal. Thus, while they form highly capable systems for ensuring adherence to treaties, they are incapable of providing objective evidence that may be shared with other less capable states, for fear of revealing the capabilities and weaknesses of the system. The inability to provide universal access to NTM data often leads to the accusation that NTMs produce politically-skewed data when introduced in multilateral forums. The use of commercial systems may suffer the slings of similar arguments if the infrastructure is not established to accept this form of factual data, and if states are not familiar with the capabilities and limitations of image interpretation. Mitigation of this challenge may entail awareness campaigns and training programmes like those conducted by the WEU Satellite Centre or those postulated in the Canadian Centre for Information, Training, and Analysis (CITA) proposal. Collective interpretation may also add objectivity to the compliance analysis process. In addition, legally-binding recognition of the use of satellite imagery may be required to diffuse the associated argument that satellite imagery is still regarded as espionage in some circles. Because of these measures, the use of commercial satellite imagery could emerge as a *lingua franca* for NACD monitoring and verification.

Another challenge for spaceborne observation assets to gain widespread uses for NACD verification is the perception that space activities are expensive and exclusive to the domains of the great powers. The fiscal challenges currently facing states demand that effective verification be achieved at ever-lower costs. To mitigate against this challenge, an awareness campaign must be waged in the United Nations and the Conference on Disarmament to show how commercial satellite imagery negates the need for a multilateral organisation to obtain dedicated satellites. NACD organisations need now only consider receiving, processing, and interpreting imagery generated by commercial satellites. What would have required billions of dollars a decade ago, only millions of dollars are now required to perform the same NACD mission.

Further challenges facing the utilisation of spaceborne imagery are the political restrictions that are often placed on the development and operations of the systems. The Landsat series of satellites was restricted to a resolution no greater than 30 metres by the United States. France offered the SPOT system with a resolution of 10 metres. With the ending of the Cold War, the Russians offered image resolutions of 2 metres. Currently, the U.S. has issued a directive that allows commercial satellite images of 1 metre resolution. Despite this competitive pressure, France will not offer any resolution below 5 metres.

Besides capability limitations, states often restrict the use of satellites during the times that the assets could be of greatest benefit. These restrictions may be temporary or permanent. For example, during the Gulf War, France may have denied access to the SPOT system to opposing interests as it was a member of the multinational coalition forces arrayed against Iraq. The U.S. may well restrict the sale of high-resolution images of Israel. To mitigate against this challenge, a multiplicity of states should place a diversity of systems in orbit to ensure that self-imposed censorship, or excessive reliance upon a single source of information does not weaken the benefits to be obtained from the utilisation of these stabilising systems in times of crisis. Acceptable international rules of conduct may need to be negotiated for the utilisation of spaceborne imagery to forestall the aspirations of states which may seek the "control" of space.

A great deal of effort has been expended in multilateral fora to obtain intrusive on-site inspections as a necessary means to provide for adequate verification of current multilateral treaty provisions. Many may see the introduction of other, less intrusive, verification measures as threatening these hard fought gains. Mitigation of this challenge will require showing that satellite observations can operate synergistically with overflight and on-site inspection measures for increased effectiveness and expanding mission roles. A complimentary mix of verification technologies with differing levels of intrusiveness may help treaty organisations seek increased capability to monitor undeclared activities. To provide a security environment suitable for the elimination of nuclear weapons, as required under Article VI of the NPT, a more rigorous verification regime will be required to detect compliance with weapons of mass destruction and their delivery systems beyond on-site inspections at declared sites. An optimal selection of verification methodologies may also be required to reduce the costly expense of on-site verification measures exercised alone. Satellite imagery should then be correctly seen as supplementing on-site inspections and not supplanting on-site inspections.

Lastly, a major challenge available to commercial satellites for NACD verification is the inherent vulnerability of these assets in space. While military systems may "harden" these assets against harm from other states of similar power, little effort is expended in commercial remote-sensing satellites to combat similar threats. Nations with commercial space assets, but without a global surveillance reach or a credible deterrence, may be wary to antagonise nations with the capability to harm satellites through clandestine efforts. To mitigate against this challenge, a legally-binding instrument should be negotiated to prohibit the intentional harming of another nation's satellites, and to prevent the weaponisation of outer space.

Conclusions

Outer space has played a key role in the verification of non-proliferation, arms control, and disarmament treaties. Initially these spaceborne systems were exclusive to the great powers, remained highly classified, and required billions of dollars of investment. Current commercialisation of space will allow every geographic region to become a player in the exploitation of spaceborne imagery for verification purposes and will reduce the costs for arms control verification to millions of dollars, as now states or organisations need only invest in the ground segment capital and the people required for the desired level of compliance monitoring. Spaceborne imagery may be used synergistically with overflight and on-site inspection methods for increased verification effectiveness. The similarity of the verification missions for the nuclear, biological, and chemical weapons of mass destruction conventions may produce increased cost-efficiencies if the monitoring mission is performed by a central verification agency. Progressive utilisation from individual states, through multinational cooperation, treaty organisation utilisation, and ultimately UN utilisation, is foreseen for high-resolution, commercial satellite imagery.

Chapter Fourteen

Summary of Results from the 1996 Workshop on the Use of Satellite Overhead Imagery in Verification

Peter Stibrany*

The primary purpose of this chapter is to provide a summary of a workshop held in March of 1996 on the use of satellite imagery in the verification of arms control agreements, concentrating specifically on the Comprehensive Test Ban Treaty (CTBT).[1]

The principal findings and conclusions of that workshop were as follows:

(1) Technically, overhead images gathered by satellites can be an important source of information relevant to verification.

(2) In particular, there are three areas in which overhead imagery can significantly improve the effectiveness and efficiency of the seismic and on-site inspection systems.

- Upon detecting a seismic event which is difficult to classify, an overhead image of the area in which the event took place could eliminate false alarms. Nuclear tests cannot be undertaken just anywhere, after all. The presence of civilian activity could rule out testing, for example. The presence of earth-moving or mining activity in a remote area may invite further investigation.
- Seismic sensing systems do not pinpoint the location of an event very accurately. The location error could be many kilometres. An analysis of overhead images of the area could very well pinpoint the exact location where on-site inspectors should go in order to verify that a given treaty was not violated. Without this additional informa-

---

* This chapter is an edited version of a presentation made to the Fourteenth Annual Ottawa NACD Verification Symposium, "Cyberspace and Outer Space: Transitional Challenges for Multilateral Verification in the 21st Century," 12–15 March 1997, Montebello, Quebec, Canada.

1. Conference on "Overhead Imagery: An Under-used Arms Control Verification Asset", 24–27 February 1996, Montreal, Quebec, Canada.

tion, inspection teams would have to cover a lot more ground in order to carry out their duties.

- An analysis of overhead imagery could identify sites where dual-use or ambiguous activity was taking place, and cue the seismic detection system to pay more attention to it. Early warning would allow site-specific calibration activities to be performed by the seismic systems, so that if there were any tests taking place in that area in the future, the seismic system could produce calibrated results sooner and with more confidence.

(3) The supply of high-quality imagery is driven by a variety of factors, including the desire on the part of state to have their own capabilities, the evolution of technology, the shrinking of government budgets leading to the search for commercial offset applications, and changes to technology transfer policy prompted by the end of the Cold War.

(4) The demand for imagery will likely be strong. Initially, security applications will probably be the most lucrative. After all, almost any state or group can now be a 'spy satellite superpower' if they have a few million dollars to spend on imagery and analysis. High quality satellite imagery will also take over a large section of the airborne imaging market. Satellite imagery will likely have more consistent quality, cover larger swaths simultaneously, present a convenient and centralised order and delivery process, and offer lower cost per unit area for many applications. Operators of airborne imaging services will likely move to providing imagery not currently obtainable from satellites, including hyperspectral imaging and active optical suites.

(5) The price of imagery will likely stay flat or reduce. It appears that, at this time, none of the factors which would tend to drive up the price of imagery are very likely to occur. These factors which could lead to higher prices are:

- the formation of a suppliers cartel;
- significant interference in image acquisition and distribution;
- absence of agreement on intellectual property issues relating to imagery;
- significant quality differentiation of image products; and,
- partitioning of the market to avoid a battle for market share among vendors.

One of the key things which may depress prices for some time is the desire of governments to offset the cost of imagery acquisition for their own purposes through the commercialisation of their imagery acquisition assets. This means that many — though not, of course, all — of the vendors of data will be able to operate at a loss, if necessary. They will not be completely commercial entities, in that sense. On the other hand, many vendors will be purely commercial ventures and will have to exist in this potentially tough environment.

(6) Overhead imagery could have many other applications besides arms control in the UN context. Besides verification and monitoring of ACD treaties and agreements, this includes support for peacekeeping operations, crop and food production monitoring, environmental protection, and disaster relief.

(7) Finally, the cost of creating and operating an effective monitoring system based on satellite overhead imagery has gone from the billions of dollars needed to set up a system twenty years ago, to only millions needed to purchase imagery and perform analysis.

It would perhaps be appropriate to talk about cyberspace and the relationship that it has with satellite-based overhead imaging. I propose to enlarge the meaning of the word "cyberspace" to include all electronic media: television, radio, telephone, as well as the various guises of the Internet. This definition comes close to the one for "electric media" as used by Marshall McLuhan some time ago.

Much has been written about the characteristics of this new media sphere and all the information which it can transmit. However, of interest to me, and I believe of relevance to the discussion of how satellites become part of this media sphere, is the information which is not readily represented or transmitted by "cyberspace." There is no handy representation of distance, for example. A voice on the telephone, a picture on television, or a page on the World Wide Web are not different depending upon their geographic location. Physical distance, therefore, finds no direct expression in cyberspace. For that matter, the whole process of travel, including formalities such as visas, passports, inoculations, and so on, largely evaporate.

In the world of business, this property of cyberspace, to remove the distance – both geographical and procedural – between two end points, has been called "disintermediation." Any procedure or organisation that does not add significant value to the process in which it exists is in danger of being circumvented, its small contribution taken up by others. I believe that "disintermediation" will cause extensive structural change to any sector of endeavor in which cyberspace is applied. This includes arms control verification.

The Internet has been used as a vehicle for transmitting remotely sensed imagery from the very first moments it was capable of doing so. There is a World Wide Web phenomenon called "web cams" referring to the use of digital cameras to feed "live" imagery to whomever wants to see it. Initially, university students used this technology to give them access to camera images of their coffee pot so that they could check that it was full before taking the long trek to where it was physically to get a cup of coffee. The Internet coffee pot was followed quickly by the Internet Coke machine and so on. Now, hundreds of cameras feed information to Internet users. Perhaps the most popular source of such pictures are the various weather satellites which allow a person almost anywhere to have a satellite picture showing the weather on their continent almost in real time.

In the language of cyberspace, satellites are almost the same as "web cams." There are certainly differences though, and the following issues must be considered:

- Satellite imagery is for sale. Most companies that sell satellite imagery do so through traditional means such as telephone, fax, and so on. Within a few years, it is very likely that one will be able to order images, pay for them, and take delivery of the result through the Internet. This will afford purchasers a new degree of anonymity.
- Satellite imagery must be interpreted. Raw satellite images are almost useless to people who are not interpreters. However, nothing prevents amateurs from becoming interpreters of satellite imagery. The cost of image manipulation and analysis tools has fallen dramatically over the years. Non-governmental organisations are becoming increasingly empowered to perform their own imagery analysis. Whereas during the Cold War, it was all but inconceivable that an amateur organisation could make meaningful statements about arms control verification issues based on their own work — the Kettering group notwithstanding — in a post Cold War environment in which media relations and cooperative measures are increasingly important to arms control, NGOs will very likely be able to take on a much more important role than they have ever done in the past.

A fairly inescapable conclusion is that, over time, the "disintermediative" power of cyberspace will begin to bring verification more directly to those who are interested in it. It would be wise to frame the verification procedures of arms control treaties to cope with this eventuality. For example, in a cooperative world, nothing would prevent the children in a high school science class from "adopting" a nuclear test site. They could order images of the site and perform image analysis on a regular basis. With the cooperation of the host government, and assuming the test site was in a fairly accessible and temperate location, this hypothetical science class could even organise a yearly field trip to perform ground truth experiments. Improbable as this scenario may now seem, I can see no technical reason why it could not come true within a year or two.

Chapter Fifteen

"93+2" Critique

Jason Cameron*

The "93+2" Programme

The "93+2" programme is the direct result of the discovery that Iraq — a State Party to the Nuclear Non-Proliferation Treaty (NPT) and subject to comprehensive IAEA safeguards — could develop a clandestine nuclear weapons programme without detection. This discovery cast doubt over the effectiveness of the IAEA's safeguards system. In an effort to remove this 'shadow of doubt', the IAEA initiated a reform of its comprehensive safeguards system. The reform process was designed "to strengthen the IAEA's comprehensive safeguards system so that its verification covers the correctness and completeness of declarations by states in order to provide credible assurances that no nuclear material is being diverted from declared nuclear activities and that no undeclared activities are taking place."[1]

The "93+2" programme was officially launched in May 1993 when the IAEA's Department of Safeguards created a special task force to lead the reform efforts by developing a coherent and strengthened comprehensive safeguards system. The reform effort was dubbed the "93+2" programme because the goal was to present to the Board of Governors a fully integrated and tested set of measures before the commencement of the April 1995 Non-Proliferation Treaty Review and Extension Conference in New York.

The development of the "93+2" programme can, however, be traced through three development paths back to as early as the mid-1980s. As I have devoted considerable efforts elsewhere to describing the development of the "93+2" programme, I will not go into detail on the subject here. The three development paths do, however, merit a brief discussion. But, first, it is also worth mentioning two important post-Iraq test cases, namely DPRK and South

---

* This chapter is an edited version of a presentation made to the Fourteenth Annual Ottawa NACD Verification Symposium, "Cyberspace and Outer Space: Transitional Challenges for Multilateral Verification in the 21st Century," 12–15 March 1997, Montebello, Quebec, Canada.

1. Richard Hooper, "The Objective of Programme '93+2'," paper presented at IAEA seminar on Verifying Nuclear Non-Proliferation and Disarmament Pledges: Future Directions for IAEA Safeguards, Vienna, Austria, 1995, 4–5.

Africa. The "93+2" programme draws considerably upon the IAEA's experiences in these two cases.

DPRK: No Cooperation + No Transparency = Non-Compliance

The DPRK signed the NPT in 1985, but did not conclude a comprehensive safeguards agreement with the IAEA until April 1992. Promptly after receiving the DPRK's initial report, the IAEA sent a team of inspectors to the DPRK to conduct *ad hoc* inspections in order to verify the correctness and completeness of the initial declaration. The data collected during the inspections revealed several inconsistencies. Some minor inconsistencies, which are normal, were cleared up immediately. But three significant anomalies remained.

- Nuclear waste, which was stored within a declared facility (the radiochemical laboratory) and from which samples were taken for (environmental sampling) analysis, proved to have an isotopic composition indicating that some plutonium should exist which had not been declared.
- Some environmental sampling analysis of the plutonium which had been declared indicated that there should be some nuclear waste which had not yet been made available for inspection, sampling, and analysis.
- The DPRK had told the IAEA that only one reprocessing campaign had been carried out in the unfinished reprocessing plant (the radiochemical laboratory) in 1990 using irradiated fuel elements from the 5MWe Experimental Power Reactor. The IAEA's analysis suggested that there had been several such reprocessing campaigns.[2]

The DPRK's declarations were at odds with the IAEA's findings and, once corroborated with intelligence information — particularly satellite image — received from its member states concerning two undeclared nuclear-related sites, the IAEA requested access to the two sites where waste from the plutonium reprocessing was thought to be stored.[3] The DPRK denied the IAEA's request. The IAEA subsequently declared the DPRK to be in violation of its comprehensive safeguards agreement because the IAEA — without the necessary access and on-site inspection — could not draw credible safeguards conclusions with respect to the diversion of declared nuclear material. The Director General reported the non-compliance to both the IAEA Board of Governors and the United Nations Security Council.

---

2. Hans Blix, Statement of IAEA Director General Regarding DPRK at Informal Briefing of UN Security Council, 6 April 1993. See also Paul Bracken, "Nuclear Weapons and State Survival in North Korea," *Survival* 35:3 (Autumn 1993), 137–152 and David Albright and Mark Hibbs, "North Korea's Plutonium Puzzle," *Bulletin of Atomic Scientists* 48:9 (December 1993), 37–40.
3. Richard Kokoski, *Technology and the Proliferation of Nuclear Weapons* (Oxford: Oxford University Press, 1995), 222–236.

The IAEA's experiences in verifying the DPRK's initial inventory underscored the value of broader access rights, environmental sampling, and intelligence information, particularly satellite imagery, in ensuring the effectiveness of the comprehensive safeguards system. These experiences were subsequently filtered into the "93+2" programme's development paths.

South Africa: Cooperation + Transparency = Correctness and Completeness

In July 1991, South Africa acceded to the NPT and in September 1991 it concluded a comprehensive safeguards agreement with the IAEA. In accordance with its obligations, South Africa submitted an initial report for IAEA verification. After several visits to South Africa, the IAEA reached a general conclusion that "it found no evidence to suggest that the declared inventory of nuclear installations and material was incomplete."[4]

Unexpectedly, on 24 March 1993, South Africa announced that it had developed and dismantled a "limited nuclear deterrent capability" involving the design and manufacture of six or seven gun-assembled highly enriched uranium (HEU) devices. Upon hearing this declaration, the IAEA — with the full cooperation of South Africa — assembled an augmented inspection team, comprising personnel assigned to carry out an assessment of completeness and nuclear weapons specialists to ascertain that all of the nuclear material involved in the programme had been recovered and placed under safeguards.[5] South Africa afforded the IAEA virtually unfettered access to all of its records, key personnel, and facilities, some of which were non-nuclear, or private, or military sites.[6]

Conducted in an atmosphere of openness and transparency, the IAEA's post-1993 inspections were able to determine "with relatively little room for speculation to the contrary," the correctness and completeness of South Africa's nuclear programme.[7] The South African case stands in stark contrast with that of the DPRK. Ironically, both offer the same valuable lesson to the IAEA and the international community: openness and transparency are the cornerstones of verifying the correctness and completeness of data pertaining

---

4. Bruno Pellaud, "Safeguards in Transition: Status, Challenges and Opportunities," *IAEA Bulletin* 36:3 (September 1994), 4.
5. Kokoski, 244.
6. Nic von Wielligh and N.E. Whiting, "Experience of an Ex (De Facto) Nuclear Weapon State with the Application of Post-Iraq Safeguards," *International Nuclear Safeguards 1994: Visions for the Future*, volume 1 (Vienna: International Atomic Energy Agency, 1994), 223.
7. Garry Dillon and Demetrius Perricos, "Verification of Completeness and Correctness of Inventory: Experience Gained in the Verification of the Inventory of South Africa's Nuclear Installations and Material," *International Nuclear Safeguards 1994: Visions for the Future*, Vol. 1, 241.

to a state's nuclear programme. This lesson resonates throughout the development of the "93+2" programme.

Development Path One: 'Piecemeal' Approach by the Board of Governors and IAEA Director General Hans Blix

The first development path can be characterised as a piecemeal approach to the issue by the Director General and the IAEA Board of Governors. Three separate measures were introduced during the February 1992 special meeting of the Board of Governors, including the following:

(1) affirmation of the IAEA's rights both to conduct special inspections and to all sources of information;

(2) early provision of design information; and,

(3) initiation of a universal reporting scheme for exports and imports of nuclear material and specified equipment and non-nuclear material.

The early provision of design information and the reporting scheme become integrated into the "93+2" programme. The important difference to note, however, is that in moving into the "93+2" programme these two measures lose their voluntary nature and become mandatory.

Development Path Two: SAGSI

Since the mid-1980s, the Standing Advisory Group on Safeguards Implementation (SAGSI) had been devising efforts to cope with increasing demands and insufficient funds. Cost effectiveness and budgetary pressures were the only driving forces behind SAGSI's activities in the 1980s. In fact, before the discoveries in Iraq, it was not considered politically acceptable for SAGSI to even contemplate activities that were designed to help the IAEA detect undeclared activities. The discoveries in Iraq, however, placed a new demand upon SAGSI — namely to strengthen the comprehensive safeguards system and improve its cost-effectiveness.

In this capacity, SAGSI made two significant contributions to the development of the "93+2" programme. First, in 1991, SAGSI proposed an Alternative Safeguards Regime.[8] The Alternative Safeguards Regime emphasised a trade-off between reduced IAEA verification in exchange for "access to any safeguards relevant data on facility operation and increased rights of access."[9] This new regime was, however, limited. First, it would only apply to the less

8. Paul Ek, "SAGSI Approaches to Strengthened IAEA Safeguards," in Tariq Rauf, ed., *Preparing for the 1995 Non-Proliferation Treaty Extension Conference: Proceedings of International Workshops on Treaty Extensions, Strengthened Safeguards, and Regional Non-Proliferation Strategies*, Issue Brief 15 (Ottawa: Canadian Centre for Global Security, 1994), 68–69.

9. *Ibid.*

sensitive aspects of the fuel cycle. Second, participation would be voluntary. Third, the Member State could unilaterally rescind the regime and return to previous agreements. While the details of the Alternative Safeguards Regime did not receive widespread support, the 'trade-off' concept forms a crucial underpinning of the "93+2" programme.

Second, SAGSI was subsequently entrusted with the initial assessment of the IAEA's comprehensive safeguards system. Its April 1993 report indicated that the IAEA's safeguards system must be strengthened so as to provide "significant confidence that no undeclared nuclear activities of proliferation relevance are being carried out in states with comprehensive safeguards agreements."[10] SAGSI's emphasis on "nuclear activities" marked a significant departure from the traditional character of the comprehensive safeguards system — namely, its preoccupation with declared nuclear material. SAGSI's report proposed three reforms:

(1) the IAEA should make better use of information already publicly available;

(2) the IAEA's Member States, and their respective intelligence organisations, should provide information and intelligence to the IAEA; and,

(3) the IAEA should develop and utilise environmental sampling techniques, which had been declassified by the United States intelligence establishment.[11]

While its first and third proposals were included in the "93+2" programme, the second suggestion regarding national intelligence information is not explicitly included.

Development Path Three: The "93+2" Task Force

Shortly after SAGSI had submitted its April 1993 report to the Director General, a new task force was created to assess SAGSI's ideas and to develop a "coherent and strengthened safeguards system, while evaluating its technical, legal and financial implications."[12] The ultimate objective of the "93+2" programme was to obtain assurances of the absence of undeclared nuclear material and activities. The task force developed a three-pronged programme to meet this objective.[13] First, broaden the IAEA's access to information and improve the analysis of that information. Second, increase the IAEA's physi-

10. *Ibid.*, 67.
11. *Ibid.*
12. Pellaud, 6.
13. See International Atomic Energy Agency, GOV/2784 — Strengthening the Effectiveness and Improving the Efficiency of the Safeguards System: A Report by the Director General (Vienna: International Atomic Energy Agency, 1995), 1–39.

cal access to sites so that its verification covers the whole site. Third, optimise present operating procedures and in-house efficiency.

The broader access to information provisions included more information on states' nuclear activities through an Expanded Declaration, the use of environmental monitoring techniques during on-site inspection, and improved analysis of all information available to the IAEA. The increased physical access to sites provisions included access needed beyond currently designated "strategic points" in safeguarded facilities to any location on the site, access as needed to other locations identified in the Expanded Declaration which are declared to be nuclear-related but not containing nuclear material, voluntary arrangements through which a state facilitates access to other locations which may be of interest to the IAEA, expanded rights to prompt access on short notice or no-notice, and "managed access" to sensitive sites under a scheme of expanded access. The optimal use of the present system provisions included greater use of unattended and remote readout equipment in lieu of some inspections, multiple-entry visas for inspectors, expanded capability for inspectors to freely communicate with headquarters, retraining of inspectors, increased cooperation with the state systems of accounting and control (SSACs), and review of the IAEA's safeguards implementation parameters.

Partition of the "93+2" Programme into Part 1 and Part 2 Measures

In June 1995, the "93+2" programme was partitioned into two parts.[14] Part 1 measures consisted of those measures which could, in the Secretariat's view, be implemented under the existing legal authority — namely INFCIRC/153. Part 2 measures consisted of those measures which were believed to require complementary legal authority. The measures can be delineated as follows:

Part 1 Measures[15]

Broader Access to Information

a. Information on the SSAC.
b. Information on past and present activities involving nuclear material.
c. Description of the nuclear fuel cycle and other activities involving nuclear material.

---

14. See International Atomic Energy Agency, GOV/2784 and GOV/2807 — Strengthening the Effectiveness and Improving the Efficiency of the Safeguards System: Proposals for a Strengthened and More Efficient Safeguards System (Vienna: International Atomic Energy Agency, 1995), 1–24.
15. See Table 1 in "A Summary of the Legal Evaluation of Measures Proposed for Strengthened and More Cost-Effective Safeguards," in IAEA, GOV/2807.

d. Information on nuclear R&D involving nuclear material at nuclear facilities and other locations containing nuclear material (LOFs).
e. The early provision of design information.
f. Environmental sampling during ad hoc inspections where the initial report or inspections carried out in connection with it indicate that nuclear material is present.
g. Environmental sampling at strategic points during routine inspections.
h. Environmental sampling during special inspections.
i. Environmental sampling for design information verification, at any location to which the Agency has been granted access to carry out design information verification.
j. Improved information analysis.

Increased Physical Access

a. Access to locations beyond strategic points in nuclear facilities or LOFs, but within the site containing the facilities or LOFs, during design information verification.
b. Unannounced/no-notice timing of inspections at strategic points.

Optimal Use of the Present System

a. Use of technological advances of safeguards technology.
b. Increased cooperation with the SSACs.
c. Changes to the safeguards implementation parameters.

Part 2 Measures[16]

Broader Access to Information

a. Information on operational activities additional to that required under INFCIRC/153.
b. Description of each of the buildings on the sites on which are located nuclear facilities, LOFs, or nuclear-related R&D.
c. Description of other locations directly related to the operation of nuclear facilities, LOFs, or nuclear-related R&D.
d. Location and status of know uranium and thorium ore deposits and mines.
e. Information on the manufacturers of major items of nuclear equipment or materials.
f. Plans for the further development of the nuclear fuel cycle.
g. Description of planned nuclear R&D activities.

Increased Physical Access

a. Access to locations beyond strategic points in nuclear facilities or LOFs, but within the sites containing such facilities or LOFS, during routine inspections.
b. Access to other locations identified in the Expanded Declaration.

---

16. *Ibid.*

c. Unannounced/no-notice timing of inspections at locations beyond strategic points within the sites containing nuclear facilities or LOFs.
d. Unannounced/no-notice timing of inspections at other locations identified in the Expanded Declaration.

Optimal Use of the Present System

a. Use of simplified procedure for designation of inspectors.

Generally speaking, Part 1 measures continue to focus on declared nuclear materials and the access provided under INFCIRC/153. Part 2 measures, on the other hand, expand IAEA's on-site verification activities and increase its access to information on nuclear activities.

During the June 1995 meeting, the Board of Governors accepted the Secretariat's intention to proceed with the implementation of Part 1 measures on the understanding that any concerns that the Member States might have would be clarified. That left the Board of Governors to consider Part 2 measures. After several unsuccessful attempts to reach consensus, in June 1996, the Board of Governors established a special safeguards committee — similar in nature and stature to the committee formed to negotiate INFCIRC/153 — to draft a model Protocol that would provide the legal authority necessary to implement Part 2 measures.[17]

The Committee on Strengthening the Effectiveness and Improving the Efficiency of the Safeguards System, established by the Board of Governors on 14 June 1996,[18] — otherwise known as Committee 24 — has met three times; in July 1996 for two days, in October 1996 and in January 1997. Although it is said to be nearing a consensus, it has yet to reach a final decision.

The Goals of the "93+2" Programme

Before embarking on a discussion of Committee 24 and the model Protocol, it is worth asking two very important questions. What is the objective of the "93+2" programme? And, how does it go about trying to achieve that objective?

Simply stated, the objective of the "93+2" programme is to strengthen the comprehensive safeguards system so that verification covers the correctness and completeness of declarations by states in order to provide credible assurances that no nuclear material is being diverted from declared activities and that no undeclared activities are taking place. This seems to imply, then, that the "93+2" programme must strive to meet three goals. First, it must

---

17. International Atomic Energy Agency, GC(40)/17 — Strengthening the Effectiveness and Improving the Efficiency of the Safeguards System: Report by the Director General to the General Conference (Vienna: International Atomic Energy Agency, 1996), 4.

18. *Ibid.*

strengthen the IAEA's ability to detect diversion of nuclear material from declared nuclear activities to undeclared activities. Second, it must empower the IAEA to detect undeclared co-located activities at declared facilities — Iraqi-style violations of the NPT. Third, it must give the IAEA the requisite tools to detect undeclared activities at undeclared locations. While the "93+2" programme goes a considerable distance to achieving the first two goals, its ability to achieve the third is somewhat unclear. Furthermore, the strength of the programme and its ability to meet these three goals clearly lies in the new model Protocol and its acceptance.

Detecting the Diversion of Nuclear Material

Part 1 measures strengthen the IAEA's ability to detect the diversion of nuclear material from declared activities to undeclared activities. Unannounced/no-notice timing of inspections at strategic points gives the IAEA an element of surprise in conducting its on-site inspection activities, making it more difficult, not to mention precarious, to conceal undeclared activities between inspections. The unattended and remote monitoring systems being tested through field trials in Sweden are providing the IAEA with near-real time accountancy reports. This information, coupled with the information provided under Part 1 — including information on the SSAC and information on past and present activities involving nuclear material and design information — gives the IAEA greater information from which to draw its conclusions. More importantly, however, these measures, taken together with the ability to conduct environmental sampling — both during routine inspections at strategic points and during design information verification — begin to expand the IAEAs view from partial to full coverage of the nuclear material in flow.

The IAEA's ability to detect diversions of nuclear material would be significantly strengthened through Part 2 measures. Unannounced and no-notice inspections are heavily dependent on the simplification of inspector designation, granting of multiple-entry visas, and visaless entry, which are all Part 2 measures. The increased physical access provisions — including access beyond strategic points and access to locations in the Expanded Declaration — and the associated environmental sampling techniques of Part 2 would also improve the IAEA's ability to detect diversions of nuclear material. Finally, the ability to follow-up location-specific inconsistencies and questions with further environmental sampling and visual inspection at that location would greatly strengthen the IAEA's ability to detect diversions of nuclear material.

Detecting Co-Located Undeclared Nuclear Activities at Declared Locations

Detecting co-located undeclared nuclear activities at declared locations is only nominally strengthened through Part 1 measures. If the undeclared nuclear activities utilise or impinge upon the declared flow of nuclear material, then unannounced inspections at strategic points and the IAEA's classical nuclear material accountancy procedures may detect the undeclared activity.

Environmental sampling at strategic points can confirm that measured nuclear signatures are consistent with the declared operations at that location and thus provide some assurance of the absence of undeclared activities. If the undeclared activity differs from the declared operations at a strategic point and the undeclared activity deposits small traces of material near a strategic point, then it is possible that swipe sampling at strategic points may detect an undeclared activity. The detection of altered isotopic ratios in uranium where only natural uranium is declared to be present, the detection of highly enriched uranium where only low enriched uranium is declared to be present, or the detection of plutonium where no plutonium has been declared, are all unambiguous indicators of undeclared activities.

Conversely, Part 2 measures would considerably strengthen the IAEA's ability to detect undeclared co-located activities at declared locations. Access beyond strategic points at declared locations and access to other locations identified in the Expanded Declaration, when coupled with the ability to conduct environmental sampling at all of these locations, would greatly strengthen the IAEA's ability to confirm that the measured nuclear signatures and observed nuclear activities are consistent with the declared operations at that location.

Detecting Undeclared Nuclear Activities at Undeclared Locations

In order to understand how the "93+2" programme attacks the problem of detecting undeclared activities at undeclared locations, it is important to understand how all of the new information is being brought together and how conclusions are going to be drawn under this new safeguards system. Senior safeguards inspectors in the three Divisions of Operations have been designated "country officers". These inspectors have particular knowledge about safeguards- and proliferation-relevant situations for the state(s) to which they have been assigned. It becomes their task, in conformity with the policies developed by the Divisions of Operations, to produce holistic country evaluations. This new approach will be significantly different from the traditional analysis conducted by the IAEA's Divisions of Operations. No longer will safeguards conclusions be drawn by cross-referencing intra-state information on nuclear material obtained on a facility-by-facility basis. The country officers are now charged with preparing an overview of the IAEA's safeguards implementation experience in that particular country, addressing each of the elements of the proliferation critical path — including source material acquisition, conversion, fuel fabrication, enrichment, power reactors, research reactors, heavy water production, reprocessing HEU and plutonium, and weaponisation — as they pertain to the country, offering an evaluation of the status of each of the elements of the critical path and recommending any follow-up action that should be pursued by the IAEA Department of Safeguards.

While the IAEA is still developing the procedures that will guide the preparation of these country evaluations, I have been offered two analogies that begin to illuminate how this process may occur. First, think of a bank audit team. It is led by a 'wise' expert who has years of experience in banking. (S)he comes to town with a team of junior auditors who fan out and randomly drill down and inspect narrow aspects of the bank's reports and records, widening their search only as needed to resolve any anomalies or questions. Having obtained the data, the team convenes and each member offers a report. After a full discussion, the leader makes a professional judgement as to compliance. In making this judgement, the leader cannot possibly look into every detail of the bank's operations — but her/his oversight nevertheless encompasses everything. In this respect her/his approach is "wide and shallow." "Shallow" in this case does not imply the inability to look deeper when questions arise, as the expert can use both her/his resources and experience in examining any particular area in greater detail, as well as enlist the support of a range of other experts as needed to enhance her/his own abilities.

The second analogy is slightly less tangible and more philosophical in its approach. In July 1945, after witnessing the first nuclear detonation at Trinty Site, Robert Oppenheimer quoted from the Bhagavad Gita, "I am become death, the destroyer." But upon closer examination, the Gita contains more innocuous commentary. From the Thirteenth Teaching of the Gita, "Knowing the Field":

> Its hands and feet reach everywhere; its head and face see in every direction; hearing everything, it remains in the world, enveloping all.
>
> Just as one sun illuminates this entire world, so the master of the field illuminates the entire field.
>
> The field denotes this body, and wise men call one who know it the field-knower.
>
> What I deem to be knowledge is knowledge of the field and its knower.

But as the Gita suggests, the IAEA can only become master of the safeguards field if its Member States empower it and equip its country officers with the tools and information necessary to illuminate the field and to produce a holistic country-wide evaluation.

The country evaluations will be reviewed by a new Information Review Committee for quality control and as an internal audit of the evaluations reached by the country officers. Generally speaking, these annual country evaluations attempt to identify any instance wherein the information might suggest the conduct of activities inconsistent with the state's declarations to the IAEA and any requisite actions.

The IAEA's ability to detect undeclared activities at undeclared locations rests, therefore, on the supposition that the nuclear fuel cycle is one of the most complex systems known and that, through the increased access to information provisions, the broader access provisions, and improved analysis, the country officer will catch an indicator of the undeclared activity. For example, an inconsistency between the state's declaration and an environmental sample, a mismatch between one country's export license and the declared activities of the recipient state, or a credible piece of open source information. This indicator would prompt the IAEA's new "wide and shallow" investigation techniques to delve "narrow and deep" into the anomaly, inconsistency, and/or question.

Given that the proliferation critical pathways analysis alone has identified about 1000 different indicators and signatures of proliferation relevance and given the amount of new information that is requested under the "93+2" programme, it seems plausible that this new system could work. A lot of the new information, however, is provided through Part 2 measures and the new model Protocol. Widespread acceptance and adherence to the Protocol is therefore crucial to the ability of the IAEA to detect undeclared activities at undeclared locations.

That said, there is another tool that remains on the shelf, ready to aid in the detection of undeclared activities at undeclared locations — namely, national intelligence, or as it is often referred to, national technical means (NTM). NTM provided by Member States, particularly high-resolution satellite imagery, could assist the IAEA's country officers in developing their country-wide evaluations. The information provided by the United States to the IAEA on the DPRK was able to corroborate and solidify the IAEA's conclusions regarding the DPRK's activities.

Another option, which could reduce the IAEA's reliance on information, is commercial satellite technology and photo surveillance. In this regard, a major study was sponsored by the safeguards support programmes of the United Kingdom and Germany that examined the application of commercial satellite technology to IAEA safeguards.[19] This study examined seven images — two SPOT panchromatic, one multispectral, and four Landsat TM — over five sites: Iraq's Al Tuwaitha nuclear research facility; the Sellafield facility in the United Kingdom; the enrichment plant at Almelo in the Netherlands; the enrichment plant at Gronau, Germany; and, the fuel fabrication plant in Hanau, Germany. The study divided the sites into declared and undeclared nuclear facilities. The following characteristics were discernible from the photos of the undeclared nuclear facilities:

- passive security in the form of more than one fence around reactors and large buildings for fuel cells in which plutonium is prepared;
- active security in the form of anti-aircraft guns and missiles and military airfields; and,
- facilities located near rivers, lakes, or reservoirs.

---

19. Bhupendra Jasani, *et. al.*, *Enhancing IAEA Safeguards Using Commercial Satellite Imagery: A Pilot Study* (Vienna: International Atomic Energy Agency, 1996).

The photos of the declared nuclear facilities were examined and the analysis revealed that the following features were observable:

- depending on the type of reactor, it may be situated near a source of water supply or it may have large cooling towers associated with it — an example of the latter is a gas cooled reactor;
- images acquired over a period could be used to assess the development and observational characteristics of a facility by detecting changes in the images;
- the use of thermal images gives an indication of the operational status of a facility;
- the T-shape buildings associated with the centrifuge enrichment facilities at Almelo and Gronau, large buildings required for housing cooling ponds for spent fuel, and reprocessing plants can be detected and recognised. However, all centrifuge facilities may not necessarily display the T-shaped building feature; and,
- power lines associated with reactors generating electricity, can be detected when there is good contrast in the image.

The study reached three tentative, and quite valid, conclusions.[20] First, there are significant potential benefits in using satellite imagery in the application of IAEA safeguards. Second, the benefits may be reaped at "relatively modest cost." Third, the transfer of the necessary technology to the IAEA does not present "insuperable difficulties."

The availability of commercial satellite imagery would greatly enhance the IAEA's ability to detect both undeclared activities at declared locations and undeclared activities at undeclared locations. As a matter of fact, satellite imagery is one of the few new safeguards techniques that directly addresses the problem of undeclared nuclear activities at undeclared locations. Disappointingly though, the IAEA and its Member States have yet to reach an appropriate *modus vivendi* on satellite imagery in particular, and NTM in general. Those Member States who have invested significant resources in developing their own national intelligence capabilities seem both unable to abate the opposition of their national intelligence establishments and reluctant to grant the IAEA the requisite in-house capabilities.

Needless to say, there is hope that the IAEA's "93+2" programme can address the problem of undeclared activities at undeclared locations, but this is dependent on the actions of its Member States. Will they accede to the Protocol? Will they grant it the in-house capability to analyse commercial satellite imagery?

---

20. *Ibid.*

Part 1 Measures: Implementation and the IAEA's Safeguards Criteria

Although the Secretariat was given the go-ahead to proceed with the implementation of Part 1 measures over eighteen months ago, numerous problems have been encountered and little progress has been made in implementing Part 1 measures on a widescale.[21] This is due, in part, to longer than anticipated education and training of safeguards inspectors in the new techniques and to a shortage of support staff. As well, consultations and discussions at various technical levels in individual states have consumed an unexpected amount of time. The principal problem in implementing Part 1 measures remains the existing 1991–1995 Safeguards Criteria. Generally speaking, the Safeguards Criteria determines the IAEA's verification goals and implementation activities in the field. Accordingly, safeguards staff must give priority to implementing the activities planned and scheduled under the existing criteria. The existing criteria already stretches the human and budgetary resources of the Department of Safeguards and its Operations Divisions, leaving little room for new activities. Until the Safeguards Criteria is amended to allow sufficient flexibility for the incorporation of Part 1 — and Part 2 — measures, this problem will persist.

Much of the "93+2" programme — particularly Part 2 measures — renders the existing safeguards criteria obsolete or, at best, disjointed from the new safeguards system. In view of the significant shift in emphasis from facility-specific evaluations to a country-wide approach, there is a need to change the existing Safeguards Criteria. This shift from facility-specific evaluations based on quantitative safeguards techniques to country-wide evaluations based on more qualitative safeguards analysis will not be easy. Few, especially the Operations Divisions, will move away from traditional techniques without a relative degree of trust in the new system. In this regard, SAGSI has begun a review of the existing safeguards criteria and appears to be adopting an "evolutionary" approach to including Part 1 measures and is trying to develop qualitative "guidelines" and "advice" to deal with those aspects of the "93+2" programme that are not easily quantifiable. Needless to say, this process is taking time.

Part 2 Measures: Committee 24, the Model Protocol, and Universality

The latest text of the model Protocol is a collection of eighteen articles, two annexes, a foreword, and a preamble. It outlines the relationship between the Protocol and existing comprehensive safeguards agreements, the provision and submission of information, complementary (expanded) access (including the scope of activities), managed access, designation of inspectors and visas,

---

21. See International Atomic Energy Agency, GOV/2863 — Strengthening the Effectiveness and Improving the Efficiency of the Safeguards System: Proposals for Implementation Under Complementary Legal Authority (Vienna: International Atomic Energy Agency, 1996), 3–6 and Dirk Schriefer, "New Safeguards Measures: Initial Implementation and Experience," *IAEA Bulletin* 38:4 (December 1996), 10.

communications systems, and the protection of confidential information, as well as, of course, entry-into-force provisions.

The Protocol is very explicit in the information requested and the timing of that information. For example, a description of nuclear fuel cycle-related R&D not involving nuclear material is requested within 180 days of entry-into-force with an update by 15 May each year, whereas information for each export of specified equipment and non-nuclear material must be provided within sixty days of the end of each quarter of each year.

It is also very explicit in its complementary access provisions and the scope of activities associated with that access. The Protocol grants the IAEA access to any place on a site and any location identified in the Protocol, provided that, if the country is unable to provide the access, every reasonable effort should be made to satisfy the IAEA's requirements. The complementary access provisions also enable the IAEA to request access to any location not identified in the Protocol — for location-specific environmental sampling. There is, however, one 'unknown' in this discussion namely, Article 4.c., which states that: "Advance notice [for complementary access] shall be in writing and shall specify reasons for access and the activities to be carried out during such access." This article is not elaborated upon and has the potential to limit, or at least stall, complementary access.

While there are other interesting happenings in the Protocol — for instance, Article 9, which allows the possibility for wide-area environmental monitoring once procedural arrangements have been made by the Board and Article 6.d. which allows the IAEA to collect environmental samples and utilisation of visual inspection to resolve location-specific questions and inconsistencies — a detailed analysis of its provisions should only be conducted once the Protocol is approved by Committee 24 and the IAEA Board of Governors.

Ironically, the main stumbling blocks in Committee 24 to concluding a final Protocol do not appear to lie in the information or complementary access provisions. They lie in connection to the Foreword and the Preamble. In particular, paragraph three of the Foreword, which requests the Director General to negotiate additional model protocols with the nuclear weapons states that would incorporate their participation in the new regime, and paragraph four which requests the Director General to negotiate additional protocols with "other states that are prepared to accept measures provided in the Protocol". Some non-nuclear weapon states feel that paragraph three is too weak and that greater pressure must be applied to the nuclear weapons states to get their participation in the new regime. Some nuclear-capable states regard paragraph four as too strong and outside the bounds of the original scope of the "93+2" programme. Needless to say, the shadow of universality looms large over the next meeting of Committee 24, which is scheduled for 2–3 April 1997.

Four scenarios, therefore, can be projected for the next meeting of Committee 24. First, and highly unlikely, universality proves divisive and scuttles any further negotiations in Committee 24. Second, and very dangerous, universality proves a problem and Committee 24 is unable to reach consensus, but passes the Protocol to the Board of Governors anyway for its special May 1997 meeting at which universality is revisited. Third, and less likely due to Director General Blix's upcoming retirement, universality proves a problem and Committee 24 agrees to meet again. Fourth, and most likely, Committee 24 reaches an acceptable wording in the Protocol on universality and reaches a consensus on the Protocol.

If this fourth scenario proves true and a Protocol is only grudgingly approved, it will be more difficult to get states to sign onto the new agreement. If they are not satisfied with the model Protocol, then they are not likely to sign it. Having posed this question in Vienna, and to various capitals, I have been informed that the residual clause to the Protocol lies in the last sentence of its entry into force provision: " ... may, at any date before this Protocol enters into force, declare that it will apply this Protocol provisionally." If this sentence is eventually interpreted by any Member State to allow selective implementation of the Protocol and of Part 2 measures, the balance of the programme, including its effectiveness, the expected trade-offs to industry and its cost-neutrality, will be destroyed. As Blix has stated time and time again, the "93+2" programme is not a menu from which member states can pick and choose according to their safeguards appetite.

Conclusion

The international safeguards community is waiting in anticipation of Committee 24's next meeting in April 1997. If the committee is able to effectively address the universality issue and reach consensus on the text of a model Protocol to implement Part 2 measures, and if the Board of Governors approves the Protocol at a special meeting in May 1997, then a significant step will have been taken toward strengthening the IAEA's ability to detect diversions of declared nuclear material and to detect undeclared activities. The IAEA's "93+2" programme relies, however, to a great extent on the information and access provisions provided under Part 2, and thus the model Protocol. Participation in the Protocol will, therefore, characterise its success. The more states that sign on to the Protocol, the more information and access that will be provided to the IAEA for it to draw its conclusions, the stronger the safeguards system. Conversely, the more states that remain outside its provisions, the weaker the system. The equation is relatively simple.

In the end, the greatest weapon in the search for undeclared activities will remain the willingness of states to demonstrate its *bona fide* nuclear non-proliferation credentials through openness and transparency. If that openness and transparency is denied and the IAEA determines that valid safeguards conclusions cannot be drawn regarding the absence of diversion of nuclear material or regarding the absence of undeclared activities, then — as in the case of the DPRK — alarm bells and whistles should be heard blowing in Vienna.

Chapter Sixteen

Light Weapons: A New Focus for Arms Control and Disarmament

David DeClerq*

The demise of the Cold War and an increasing awareness of intrastate conflict, largely ethnic in nature, has focused attention on an area of arms control hitherto little explored: namely, micro-disarmament as it applies to light weapons. It has been recognised for some time that light weapons are the primary tools in generating human casualties. Until recently, however, international arms control studies and efforts have centered almost exclusively on weapons of mass destruction and their delivery systems, and large conventional weapons systems such as tanks, armoured combat vehicles, artillery, combat aircraft, and the like.

While light weapons as an arms control and non-proliferation issue is a relatively recent phenomenon, an annotated bibliography recently published by the Department of Foreign Affairs and International Trade, Canada, notes well over 80 publications on the subject.[1] It has been the topic of some twelve UN resolutions and documents and is a significant area of study for several security research organisations and academic institutions.[2] Numerous presentations on the subject have been given in official and unofficial

---

* This chapter is an edited version of a presentation made to the Fourteenth Annual Ottawa NACD Verification Symposium, "Cyberspace and Outer Space: Transitional Challenges for Multilateral Verification in the 21st Century," 12–15 March 1997, Montebello, Quebec, Canada. This chapter draws much of its content from a study completed by the author for the Department of Foreign Affairs and International Trade entitled "Light Weapons and Micro-Disarmament." In addition, the author draws upon his experience as the Canadian representative on the UN Panel of Governmental Experts on Small Arms. The views expressed in this chapter are those of the author and do not necessarily reflect those of the Department of Foreign Affairs and International Trade or of the Government of Canada.

1. Canada, Department of Foreign Affairs and International Trade, *Small Arms and Light Weapons: An Annotated Bibliography* (Ottawa: DFAIT, November 1996).

2. See Joel Johnston, *Current Projects on Light Weapons* (London: British American Security Information Council (BASIC), February 1996), and numerous United Nations Institute for Disarmament Research (UNIDIR) publications on small arms and micro-disarmament.

fora.[3] The literature — academic and official; theoretical and applied — is attempting to grapple with a number of conceptual and operational issues concerning light weapons proliferation. Only a few of these issues will be discussed in this chapter.

Probably the first significant international impetus to more seriously addressing the issue was in January 1995 when the Secretary General of the United Nations stressed the need for "practical disarmament in the context of the conflicts the United Nations is actually dealing with and of the weapons, most of them light weapons, that are actually killing people in the hundreds of thousands."[4] It was within this context that the term micro-disarmament was first coined. For my purposes, I define micro-disarmament as the collection and subsequent disposal of light weapons, normally through destruction, used in internal political conflict. This would not necessarily exclude the inclusion of ceilings on light weapons holdings nor the application of micro-disarmament to interstate conflict.

Arms control has always sought to deal with controlling the weapons endemic to interstate conflict.[5] Arms control has generally been defined in military-strategic terms and thus deals with weapons intended for use by the military forces of the world on behalf of the governments or states that they represent. Domestic lawlessness, on the other hand, has by and large been viewed as the legislative, police, and judicial responsibility of individual states. This does not mean that international cooperation is not applicable to lawlessness or criminal activity where it involves light weapons, or anything else for that matter. But it equally does not necessarily imply that light weapons used in criminal activities are an arms control issue in the classical sense. However, much of the literature on light weapons and small arms as a proliferation issue tends not to treat the issues of law and order and international and regional security separately. Law and order, or criminality, is sometimes intermingled with organised political violence in the context of light weapons proliferation, which thus contributes to the complexity of the issue as an arms control or proliferation matter. One must ask the question, where does 'arms control' or 'non-proliferation' within the realm of international security end, and where does 'firearms regulation and enforcement' within the realm of domestic law and order begin?[6] If the definition remains ill-defined, there will

3. Several unpublished papers and presentations were given at, for example, The Pretoria Workshop on Small Arms and Light Weapons in Southern Africa, September 1996.

4. Boutros Boutros-Ghali, *Supplement to An Agenda for Peace: Position Paper of the Secretary-General on the Occasion of the 50th Anniversary of the UN*, UN General Assembly Document A/50/60-S/1995/1, 3 January 1995.

5. In this chapter, arms control and non-proliferation will be used interchangeably, as both concepts relate to the issue of light weapons proliferation.

6. See David Mutimer, "Non-Proliferation and International Security," in *Control But Verify: Verification and the New Non-Proliferation Agenda* (Toronto: York Centre for International and Strategic Studies, 1994), 10. Mutimer defines proliferation as "the

be regime confusion, differing objectives, and many more actors involved which could slow or undermine cooperation and eventually constrain any potential successes in controlling light weapons proliferation. Some might argue that this is a humanitarian issue — not a security issue — and point to landmines as a worthy example of arms control seeking bans on humanitarian grounds. The humanitarian aspect is not the question — arguably poison gas, 'dum-dum' bullets, and nuclear weapons were humanitarian issues and not just international security issues. What is clear about those weapons and landmines is that they were designed for and used by the armed forces of states in pursuit of a state's interest, or were used by organised armed political groups *against* the state and were thus legitimate security concerns.

Another issue of some import concerns defining light weapons. Should all firearms be included — that is those clearly designed for civilian use such as shotguns and hunting rifles — or should weapons clearly designed for military use be the only target of international ACD? There are good reasons to restrict any arms control initiatives to military light weapons only — if for no other reason than to reduce the complexity of the issue and to avoid treading into the realm of domestic firearms legislation where various lobby groups and governments could severely hamper any attempts at addressing the issue.

One of the key questions to ask with regard to problems and issues dealing with light weapons concerns whether to focus on misuse from a supply-side, a demand-side, or both. Much of the literature analyses light weapons proliferation through the supply-side spectrum of diffusion. In a comprehensive analysis of supply sources for light weapons, Michael Klare compartmentalises transfers into nine categories: government grants; government sales or transfers; commercial sales; technology transfers; covert arms deliveries; government gifts to allied paramilitary groups; the black market; theft; and, exchanges between insurgent and criminal organisations.[7] They can be grouped as follows.

Government. Direct government involvement includes grants and transfers at little or no cost, on a state-to-state basis, and is a relic of the Cold War. Government sales are transfers of surplus weapons or weapons from government-owned factories. The potential impact of surplus weapons as a contributing factor in light weapons proliferation cannot be underestimated, whether

---

diffusion of weapons, associated technologies and expertise that produces an adverse effect on local, regional or global security and stability." This definition implies military security, but it can mean security in the broadest sense.

7. See Michael T. Klare, "Light Weapons Diffusion and Global Violence in the Post-Cold War Era," in *Light Weapons and International Security* (Nirankari Colony, Delhi: Pugwash Conferences on Science and World Affairs, British American Security Information Council, Indian Pugwash Society, and Institute for Defence Studies and Analysis, 1995), 8.

they are deemed to be government or commercial, particularly in view of the number of weapons that are currently surplus.[8]

Commercial Sales. Commercial sales, according to Klare, predominate in all but a few states. However, to suggest that there is no link to governments is misleading. All states, particularly advanced states, are generally aware of these transactions, as most have some form of import/export controls. As Jasit Singh points out, "the state is the facilitator in the spread of small arms in the overwhelming majority of cases around the world."[9]

Technology Transfer. Klare points out that technology transfers aid in domestic arms production in twenty-two developing states, where 66 out of 146 weapons are produced under licence.[10] In some instances the prospect, or actual implementation of embargos from previous supplier states has encouraged this type of diffusion. In this regard, embargoes can be a two-edged sword.[11] Moreover, technology transfers, like commercial sales, are not normally done without the participation of governments. It should be noted that reverse engineering — technology theft — is relatively easy to accomplish, particularly with light unsophisticated weapons.

Quasi-Legitimate Transfers. This is the term that I have coined for two of Klare's diffusion sources. Covert arms deliveries by governments are what Klare considers the most destabilising as they are going to non-state actors with few controls. Afghanistan, Mozambique, Angola, Cambodia, and Nicaragua are noted as the worst examples. As well, government gifts to allied paramilitary groups run a close second to covert deliveries by governments.[12] The most damning example given is the Hutu-dominated government in

---

8. In 1996, a multinational inspection team carried out a CFE Treaty inspection in Belarus. As permitted by the Treaty, several large warehouses within a declared site were inspected. According to a Canadian inspector who was present, in the first warehouse, the inspectors observed long, rectangular wooden crates. External markings indicated the presence of small arms, possibly twelve to fourteen per crate. There were at least ten other warehouses filled in a similar fashion. Some crates were open and the presence of AK-47, AKM, AKS, and Dragunov sniper rifles was confirmed. At least 100 crates of Makarov semi-automatic pistols were also observed. It was estimated that there were probably over a quarter of a million firearms in all. The stockpiled firearms are believed to have been a legacy of the Soviet era. At present, the firearms belong to the Belorussians, who have no use for them except as a saleable commodity. Rail and trucking facilities were in direct proximity to the warehouses. The local Belorussians jokingly offered to sell or trade the firearms to members of the team.

9. Jasit, Singh, ed., *Light Weapons and International Security* (Nirankari Colony, Delhi: Pugwash Conferences on Science and World Affairs, British American Security Information Council, Indian Pugwash Society, and Institute for Defence Studies and Analysis, 1995), 154.

10. Klare, 12.

11. Author's discussion with Pakistani officials and military officers, Pakistan Ordnance Factories, December 1977. Pakistani officials cited the arms embargo during the 1965 Indo-Pakistani War as the incentive to become self-sufficient, particularly in munitions.

12. Klare, 13.

Rwanda supplying arms to the Interhamwe Militia. Another example is Yugoslavia providing arms to Serbs outside its jurisdiction in the early days of the Yugoslavian breakup. Again, these are state-sanctioned transfers.

Illicit Transfers. Finally, there are black market arms transfers, which, according to Klare, are now more important as a result of UN embargoes and the increase in non-state actors as belligerents.[13] On the other hand, Karp suggests that the black market as a major source of arms is a myth, as the trade is too expensive for most sub-state actors and too risky for individual black marketeers.[14] Klare's examples of the IRA and drug cartels certainly indicate that only well-financed organisations can participate in the black market. It appears, however, that both authors may be making suppositions because so little is known about the black market — it is an area that merits more study.[15] Theft, another form of illicit transfer, stems mostly from state sources such as police and military armouries. Exchanges between insurgent and criminal organisations constitute the final diffusion source. Here, examples would include the Colombian drug cartels, insurgent groups, and drug and precious stone cartels in Southeast Asia.

Demand

Intrastate conflict brought on by the political, economic, and social conditions within a country often feeds the hostilities of ethnic, religious, language, and extreme ideology — particularly where democratic institutions are lacking or are unable to address the issues. Malevolent political conflict, in turn, can initiate violence which can create a demand for a variety of light weapons; or, where they are already available in large numbers, increase the level and intensity of hostilities, exacerbating the situation even more. Often, in conjunction with political violence, and occasionally even in its absence, criminal activity can threaten the domestic security of a state. Both political and criminal violence may in turn effect the security and safety of a state's citizens and those of neighbouring states.

Violent political situations and uncontrolled criminal activities can fuel an even greater demand by the security forces of a state for more or better light weapons to combat the problem. In turn, where security forces are unable or unwilling to provide protection to the citizens of the state, then 'law abiding citizens' will also demand light weapons, in order to provide for their own protection. Even if a solution is reached through negotiation, containment, or violence/counter violence, the instruments of violence remain to arm another conflict.

Almost all the literature on light weapons proliferation acknowledges that the underlying issue of violence is demand *not* supply. The first regional work-

---

13. *Ibid.*, 14.
14. Aaron Karp, "Arming Ethnic Conflict," *Arms Control Today* (September 1993), 8–13.
15. R.T. Naylor, "The Structure and Operation of the Modern Arms Black Market," in *Lethal Commerce: The Global Trade in Small Arms and Light Weapons* (Serial No. 502), 1995, 47–57.

shop convened by the UN Panel of Governmental Experts on Small Arms identified five major issues relevant to light weapons proliferation. There was a strong demand element in three of the major issues: an absence of adequate avenues for gainful employment; restricted economic opportunities and rising expectations; and, issues in part based on ethnic affinity. By the same token, the Regional Workshop in El Salvador emphasised many of the same problems as sources of demand.

Recent studies show that 91 of 96 armed conflicts accounted for in the last seven years (1989–1995) were intrastate conflicts.[16] In terms of the ethnic component to these wars, the catalyst was, according to one source, "collective fears of the future."[17] In essence, conflict occurs because the state becomes weak and is perceived as unable to control events that effect the welfare and security of its citizens. Frequently, ethnic differences are seized upon by agitators to explain or exploit this lack of security. It is partially within the context of 'fear for the future' that the demand for, and violent use of light weapons should be seen.

There appears to be a dilemma in some of the literature in addressing the demand side of the equation in light weapons proliferation. It centres around the idea of 'legal and legitimate', 'just and unjust', 'order and disorder'. These issues are difficult to address as they are often subjective and sometimes involve choices in which neither solution is particularly attractive. Is the legal government the legitimate government? Is the sub-national group involved in a just or unjust conflict and is order — an absence of armed conflict — more important than disorder — armed conflict — even if it means that a just war against an illegitimate government with the aim of correcting social injustices must be terminated? In other words, when do the ends justify the means?

While there are other issues regarding light weapons proliferation not yet discussed, most are subsets of supply and demand. Encompassing both supply and demand is what Singh calls norms and inhibitions — in essence, education, public dialogue, and transparency.[18] This is an area where advocacy groups, NGOs, the media, governments, and regional and international bodies can all play a role in stigmatising violence and the tools of violence as an illegitimate means to an end. It is a popular solution in the literature, and one that is relatively easy to implement in comparison with many supply- and demand-side solutions. Its effectiveness, however, can only be seen as slow, long-term and, in some instances, problematic.

---

16. Peter Wallensteen and Margaret Sollenberg, "The End of International War: Armed Conflict 1989–95," *Journal of Peace Research* 33:3 (August 1996), 353–370.
17. David Lake and Donald Rothchild, "Containing Fear: The Origins and Management of Ethnic Conflict," *International Security* 21: 2 (Fall 1996), 41–75.
18. Singh, xi.

The UN Panel of Governmental Experts on Small Arms

The roots of this panel can be traced to several UN sources.[19] UN Resolution 49/75M addresses illicit arms trafficking and Annex 1 of UN Document A/51/42 provides "Guidelines for International Arms Transfers." The Panel of Governmental Experts on Small Arms was formed on the basis of UNGA Resolution 50/70B on small arms. The Panel is to prepare a report on behalf of the UN Secretary General, paying particular attention to the role of the UN and the complementary role of regional organisations, covering the following areas:

- the type of small arms and light weapons actually being used in conflicts being dealt with by the UN;
- the nature and causes of the excessive and destabilising accumulation and transfer of small arms and light weapons, including their illicit production and trade; and,
- the ways and means to prevent and reduce the excessive and destabilising accumulation and transfer of small arms and light weapons — in particular as they cause or exacerbate conflict.

To date, the panel has worked toward consensus in the first two areas. The more difficult challenge is the third area concerning prevention.

Canadian proposals focus specifically on a well defined area — conflicts dealt with by the UN in the context of peace operations. The formulation of these proposals was based on a review of literature and discussions with practitioners analysing numerous peace operations in which a micro-disarmament component was present, implied, or attempted. Most, but not all, of the operations in question addressed the aftermath of a civil war or insurgency. These involved the parties signing a peace accord and voluntarily surrendering their weapons as part of a reinsertion and reintegration back into the political, economic, and social fabric of the community.

Outlined below are a sample of some of the many observations on micro-disarmament carried out under UN Peacekeeping operations. These comments deal primarily with Mozambique but, in many cases, apply across a spectrum of UN operations — i.e., El Salvador, Somalia, and the former Yugaslavia.

- Neither party trusted each other, thus information provided to the UN was lacking or inaccurate.
- Neither party had sufficient control over its own military resources.
- There was insufficient international funding to implement the reintegration support scheme for demobilised combatants.

---

19. See UN General Assembly, *Measures to Curb the Illicit Transfer and Use of Conventional Arms*, UN Resolution 49/75M, 15 December, 1994 and UN General Assembly, *Report of the Disarmament Commission*, UN General Assembly Document A/51/42, 1996.

- Implementation of the demobilisation process in a timely fashion was not possible because the ceasefire commission was not ready, there was little trust and conciliation, and, assembly sites were too few, not developed, and not accessible to all.
- There were delays in demobilisation due, in part, because too few combatants wanted to join the new army.
- Registered arms in assembly areas were not secured.
- Most munitions handed over were of poor quality — the UN suspected that better munitions were held back.
- Not every soldier demobilised had a weapon.
- Belligerent parties sometimes refused UN authority to collect and disable weapons at so-called non-assembly areas.
- Verification of weapons was to take place after demobilisation, but demobilisation was slow and parties were unenthusiastic about verification.
- The General Peace Agreement was too non-specific concerning disarmament and arms destruction, allowing parties to stall and circumvent.
- The arms collected, rather than being destroyed, were handed over to the new government and military. These far exceed their requirement and most of what was collected may well be back in circulation.
- Inability of the UN to guarantee the safety of those who voluntarily handed over firearms.
- Absence of a reward or incentive scheme for complying with disarmament and no penalties for non-compliance.
- Possession of firearms was seen as a cultural, social, and personal security issue. Many saw no logic in handing over personal firearms.
- The UN was not always seen as neutral.

Based on these observations, it seems clear that micro-disarmament must be approached in a holistic and systematic manner. First, security of person has to be guaranteed. As well, a reasonable degree of political and economic security is required. Thus, as always, part of the solution is very much dependent upon resources — i.e., UN peacekeeping resources and financial assistance to the parties. If the demand-side is properly addressed the supply-side enforcement becomes easier to implement. As can be seen by the examples cited above, it would appear that the systematic approach to micro-disarmament was lacking.

To this end, Canada has made a number of recommendations to the UN Panel. The first step to micro-disarmament within the context of UN peace operations is to ensure that peace negotiations taking place in a given state — whether brokered by a UN facilitator, a guarantor state or states, other third parties, or various groups within the state — put micro-disarmament on the table as part of the negotiations. A set of guiding principals should be developed for negotiators outlining the benefits of micro-disarmament. Principles to be applied could include such concepts as confidence and security-building measures (CSBMs), good faith, equality, sufficiency, and public safety. They should also put forth suggestions for implementation, such as guidelines for assembly areas, verification, control, and destruction of light weapons. Although it is difficult to guarantee funding at this stage, the offer of financial assistance, commensurate with the quantity and quality of light weapons handed over for destruction, is a necessary consideration. Assistance could be a combination of resettlement grants and construction aid, demobilisation bonuses, and assistance in establishing small business ventures for employment purposes on both an individual and collective basis. The social, economic and cultural dimensions of the region in question must be taken into account in designing assistance packages.

Although it may be difficult to obtain agreement with all former belligerents, there should be a reward and penalty factor for compliance or noncompliance within the micro-disarmament component of a peace agreement. This may include a bonus for the number of weapons handed in over a certain number, or denial of assistance if, for example, arms caches are discovered that have not been reported or if certain elements refuse to cooperate in either searching for or handing over weapons.

Once a micro-disarmament agreement has been secured, care must be taken that the mandate accurately reflects the requirements of the peace agreement. Particular attention has to be given to financial and force resource allocation. There must be adequate time between the signing of a peace accord and force deployment for purposes of disarmament, to permit the belligerents to organize their forces and prepare for demobilisation and the handing over of weapons. At the same time, a balance will have to be struck to ensure full implementation in as short a time as possible to prevent 'back-sliding' on any agreement.

The micro-disarmament component of the mandate probably requires a central coordinating staff within the mission headquarters and perhaps a disarmament staff at the Directorate of Peacekeeping Operations in New York — leaving the implementation to sector and contingent commanders to carry out, in accordance with their own interpretations, may cause uneven application and indifferent results. Just as each headquarters has an operations officer, a logistics officer, an information officer, and, in some cases, even personnel designated as economics and agricultural officers, so too should it have a disarmament officer. The staff should be made directly responsible for coordinating all aspects of the disarmament component of an agreement. This is not to suggest that commanders, contingent or sector, be bypassed. Responsibility would include the development and application of standard operating procedures and the organisation of weapons collection points, verification, security, and destruction. In conjunction with a proper staff system, disarmament teams could be formed with sole responsibility for the micro-disarmament aspects of the agreement. These would be officers and NCOs, knowledgeable in weapons identification, verification, and destruction. Depending on the mandate of the mission, they could work alone or in conjunction with sector and/or contingent commanders and their forces.

If a 'buy back' or reward system is used to encourage the surrender of weapons, it should be a sliding scale of reward directly related to the lethality of the firearm concerned. As was done in Haiti, automatic weapons in good condition would warrant a much greater reward than a broken shotgun.

Another method, which would require cooperation from all sides, would be to decide during the peace negotiations what the minimum requirement for surrender of weapons would be before a reward system was implemented. Collective assistance bonuses could be provided for certain benchmarks over that number. Penalties may also be incorporated within the terms of the peace agreement by stating that after a certain date anyone possessing a firearm who is not authourised to do so will be subject to certain penalties by the state. Presumably, this would be after an election to ensure even application.

It might also be advantageous to build into a peace agreement and mandate the right to inspect certain areas or installations, much as is done in both the CFE Treaty and Vienna Document. If properly addressed, this might overcome the tendency to refuse the UN access to certain areas — as has been done on previous UN missions — which, in turn, might encourage more openness on the part of conflicting parties.

These recommendations are by no means complete. However, they do highlight some of the more important considerations if micro-disarmament is to be an effective component of UN peace operations. It is these recommendations that Canada has put forward to the UN Panel for inclusion in the final report. If for some reason, they are 'watered down' or rejected, there would seem to be little hope for substantive progress on this issue from a UN perspective. There do seem to be encouraging attempts to address the issue at the regional and national levels, based on some of the feedback from regional workshops dealing with the issue.

There are many other areas of this rather daunting subject, such as domestic firearms regulation and illicit weapons transfers. Much of the academic literature addresses the merits of a "light weapons registry". Light weapons proliferation is an issue where rapid or even moderate progress is unlikely. However, incremental steps can be taken in selected areas to lessen the impact of destabilising surpluses of light weapons.

Chapter Seventeen

The Russian Crisis and Prospects for Arms Control

Sergei Plekhanov*

The past ten years may well go down in history as 'The Great Decade of Arms Control.' Since the breakthrough at the U.S.-Soviet INF talks in 1987, the world has seen a remarkable turnaround from the Cold War arms race — which had continued for half a century — to real, if not comprehensive, arms reductions. A system of bilateral and multilateral arms controls, historically unprecedented in scope and effectiveness, has been created. Their implementation relies on a global web of national and international institutions.

It might be premature to think that the momentum of the Great Decade is now exhausted — we shall see in the next few years. But the signs of trouble are visible. The ABM Treaty, regarded for a quarter century as the cornerstone of arms control, is increasingly tenuous, under pressure from new American defence programmes and from politicians who dismiss it as obsolete. START-II is bogged down in the Russian State Duma, which links its ratification with NATO enlargement *and* the fate of the ABM Treaty. The CFE is being tested for durability at the Vienna adaptation talks, which promise to be long and arduous. The Chemical Weapons Convention (CWC) was passed by the U.S. Senate only after the Clinton Administration made concessions to Senator Helms, which have created new obstacles for arms control in other areas; and the Russian Duma has refused to ratify it on the grounds that Russia does not have the funds to destroy its huge CW stockpile. The Nuclear Non-Proliferation Treaty (NPT) was extended in a weaker form than was desirable.

It could hardly have been otherwise, of course, since the unique historical conjuncture which made the strides in arms control possible since the late 1980s is now a thing of the past. A new combination of factors is at work, and one of the major areas of change is Russia.

---

* This chapter is an edited version of a presentation made to the Fourteenth Annual Ottawa NACD Verification Symposium, "Cyberspace and Outer Space: Transitional Challenges for Multilateral Verification in the 21st Century," 12–15 March 1997, Montebello, Quebec, Canada.

The Soviet Union's strategic disengagement from the Cold War was a major, if not *the* major, factor for the past decade's progress in arms control. Gorbachev and his group of reformers in the Soviet leadership undertook the disengagement out of a conviction that continuation of the Cold War was contrary to the national interests of the Soviet Union. Soviet reformers knew only too well that serious changes in the Soviet system were impossible under the conditions of a global ideological and military conflict with the West. An end to the arms race would help free up needed resources; a less conflictual relationship with the West would help weaken the bureaucratic stranglehold on society; the Soviet economy would benefit greatly from integration with the world economy; a renunciation of the Brezhnev Doctrine would bring about more normal relations with East European countries, and thus help Russia overcome its isolation from Europe.

As Gorbachev pushed his reform agenda, Soviet conservatives were mollified by the fact that he proceeded from a position of strength, inherited from his predecessors. He could claim that the USSR was secure because for decades it had invested more than enough in military power — yet at the same time argue that a de-emphasis on military power was now imperative due to changed world realities. The Soviet Union was no longer afraid of the West; it wanted to cooperate with it across the board. In other words, the West's original partners in the Great Decade were not a cornered bunch of 'losers,' but reasonably self-confident people who thought they had considerable room for maneouvre and were determined to do bold things in their country's best interests. The steps usually seen as Soviet concessions to the West in the Great Decade — the INF deal which eliminated four Soviet warheads for every one by the U.S., the decision to let Eastern Europe leave the Soviet bloc and the withdrawal of Soviet troops from the region, Moscow's acceptance of German unification, the agreement to eliminate heavy MIRVed ICBMs, etc. — were regarded by Kremlin reformers as jettisoning expendable baggage (obsolete kinds of national power) and investing in the kinds of power perceived as more durable, flexible, and modern.

The reasons why it did not quite work out this way are another story, but obviously the end of the Great Decade finds quite a different situation in Moscow. Gorbachev is a sad and vilified figure in post-Communist Moscow, desperate to be understood. His successors in the Kremlin are profoundly uncertain about how to proceed. Indeed, it may yet turn out that it was easier to do productive arms control with the Soviet superpower than it is with its downsized and frayed Russian remnant.

Originally, of course, Yeltsin was even better than Gorbachev as a partner in arms control — the new Russian state which emerged from the Soviet collapse simply completed the strategic disengagement. The Russian Federation accommodated the West on nearly all outstanding issues. Following the logic of Gorbachev's 'New Political Thinking,' Yeltsin's reformers saw a strong synergy between disarmament and domestic reform. It seemed that a would-be capitalist and democratic Russia had even less reason to fear the West than a reforming USSR. Moreover, a policy of Westernization logically *presumed* a thorough reconciliation with the West. Recently, however, different perceptions of the foreign-domestic nexus have been gaining ground among Russian elites. While the troubled transition to capitalism is eroding pro-Western sentiment in Russia, Moscow is evolving toward a more nationalist-oriented foreign policy which would expect less from, and rely less upon, cooperation with the West. Two aspects of the Russian situation are especially important to note in this context: a *transition crisis* and a *failing state*.

Dimensions of the Crisis

The essence of the transition crisis is that the policies of post-Communist reforms have entailed exorbitant costs for the Russian economy and society, while empowering and enriching a very narrow group of elites. Six years ago, when the radical market reforms were launched, their rationale was that the standard neoliberal remedy of 'Three Zations' — liberalization, stabilization, privatization — offered the only way out of the Soviet economic deadlock. But after six years, the Russian economy is still in a depression, its technological base continues to fall apart, and the market thrives at the expense of production. A vast majority of Russians have seen their savings evaporate and incomes decline to subsistence minimum or below. The impact of austerity budgets on Russian society, which for generations relied on the state for the provision of social services, has been devastating, reflected in a demographically rare peacetime decline of male life expectancy from 63 to 58 years since 1991. Even if the promised recovery should start in 1998, as both the Russian Government and the IMF expect, Russia will attempt to work out a different set of public policies, some of which the West is likely to object to.

The fact that the entity called the Russian Federation is a failing — though not failed — state is one of the manifestations of the transition crisis. The Russian Government rests on a very weak constitutional foundation — rule of law remains a distant ideal, the state has trouble assuring its territorial integrity, collecting its taxes, and feeding its soldiers. Indeed, it is hard to say what the Russian government *can* do more or less effectively.[1] Elite consensus is achieved through an authoritarian, tsar-like presidency; the parliament, designed by Yeltsin as a much weaker branch of authority, is an aggrieved and rebellious body anxious to undermine the presidency; the rules of elite conduct are minimal; the threat of a *coup d'etat* hangs in the air, mitigated largely

---

1. In a remarkably candid speech to the Russian parliament in March 1997, President Yeltsin admitted, "Having created a new political system, we have failed to install its control tools and are still incapable of governing in a new manner. Having left the old shore, we are still flailing about in a stream of problems. It carries us on and prevents us from getting to the other shore." National News Service, "Situation in Russia," 17 March 1997.

by the opposition's inability to offer realistic alternatives to the Yeltsin regime.

The transition crisis and the failing state have two kinds of consequences for arms control. First, there are *side effects* of the Russian crisis — dangerous things which happen despite Moscow's intentions and due to its inefficiency and corruption. Second, there are crisis-generated *political trends* which affect Russian foreign policy in ways detrimental to arms control.

Side Effects

The transition crisis has created difficulties for the implementation of existing arms control agreements and the maintenance of the Russian military arsenal. In a sense, arms control has become one of the victims of the monetarist logic of post-Communist economic reforms, as drastic budgetary cuts undermined the Russian state's ability to fulfil its international obligations, take proper care of its armed forces, and introduce long-promised military reforms. Financial difficulties were experienced by Ukraine, Belarus, and Kazakhstan as they set out to dismantle the nuclear weapons which remained on their soil after the Soviet collapse. But for Russia, which adopted a much tougher monetarist policy and, at the same time, assumed full responsibility for the Soviet nuclear arsenal and the bulk of Soviet conventional forces, the challenge has been enormous.

According to a recent French Defence Ministry report, the Russian Government does not know the exact number of nuclear arms in its arsenal and is destroying fewer weapons than it should.[2] The CTR (Nunn-Lugar) legislation adopted by the U.S. Congress in 1992 has been only partially helpful, having provided a total of $75 million in assistance to Russian nuclear arms reductions. Financial problems have been cited in connection with dismantling of nuclear weapons, disposal of treaty limited equipment (TLE) withdrawn beyond the Urals under the CFE Treaty, ratification of START-II involving a costly restructuring of Russia's strategic-arsenal, and of the Chemical Weapons Convention.

The problems of maintenance of Russia's arsenal, especially its WMD components, have been receiving more attention recently. The issue was dramatised in February 1997, when Defence Minister Igor Rodionov warned that as a result of inadequate financing, control of Russia's nuclear weapons — both those stored after dismantling and those still operational — might be lost in the coming years.[3] Although some observers associated this statement with the military's fight for more budget appropriations, most experts agree

---

2. The report estimates the number of Russian strategic weapons at 6,650 and tactical weapons at 18,000–20,000. Because of insufficient funding, some of the weapons have been made inoperative instead of being dismantled. *RFE/RL Newsline*, 5 May 1997.

3. *Reuters*, 7 February 1997.

that the warning should be taken very seriously. According to Bruce Blair of the Brookings Institution, a leading authority on nuclear arms, over 3,000 tactical nuclear weapons, small and susceptible to theft, are still in the field under questionable safeguards. But a much bigger problem, according to Blair, involves strategic forces:

> Although there is supposed to be preferential allocation of resources to the nuclear units, that really hasn't happened, and as a consequence they're suffering the same kinds of hardships as everyone else in the military, ranging from not being paid in months to not having resources to train and operate their systems. There is also a general deterioration of the infrastructure of nuclear control and early warning systems as a result of the budget austerity and the loss of key personnel from the design bureaus ...[4]

Since both U.S. and Russian strategic forces remain on alert, just as in the Cold War days, truly hair-raising possibilities come to mind.[5]

The *proliferation* hazards accompanying the Russian transition crisis arise from two types of situations. The first type involves *private* activities of military personnel or defence industry employees who may be tempted to sell weapons, military technologies, and their own skills to the highest bidder, which may be an organised crime group or a foreign government. Lack of a stable and effective system of government in Russia, combined with widespread corruption, makes such activities possible — the destitute conditions of the army and defence industries, creates strong incentives for them.

The second type involves *official* actions of the Russian Government, motivated primarily by commercial consideration, to promote arms and arms-related exports which may carry proliferation risks. Oil and arms are the two main export items Russia has, and fostering arms exports has become a major goal of the Russian Government as it struggles with the economic slump and the budget deficit. Of course, Russia is party to NPT, MTCR, and Wassenaar, and officially committed to non-proliferation. But some deals, like the sale of nuclear reactors to Iran, have aroused proliferation concerns, and, whenever they are voiced, the tendency in Moscow is to suspect the West of double standards and a desire to keep Russia out of lucrative markets. But the issue goes deeper than what is and what is not allowed under a particular agreement: the fact that Russia is reviving the Soviet practice of large-scale arms exports

---

4. *Voice of America*, 16 April 1997.

5. Stephen Rosenfeld, for example, outlines "the nightmare scenario of an accidental Russian launch, that, under prevailing American doctrine, would trigger a first strike or quick second strike aimed at a full range of Russian military targets, including command and control, nuclear and conventional forces, and the capacity to wage war. Since Russian missiles are similarly postured, the Henry L. Stimson Center reports, 'an accidental Russian launch could trigger an overwhelming and unnecessary disaster for both countries.'" Stephen Rosenfeld, "Wake Up – The Nightmare's Not Over," *Washington Post*, 31 January 1997.

in order to rebuild itself as a market economy raises big questions about post-Communist reform strategies — and also about the international order which puts a premium on the arms trade.

Brain drain from the Russian military-industrial complex is seen by some analysts as one of the most dangerous proliferation problems. "Foreign qualified scientists, technicians, and engineers can be more important to a clandestine nuclear weapons program than generally assumed or acknowledged. Their importance does not diminish the danger of illicit nuclear materials trade, but weapons-usable material alone will not necessarily propel either a nascent or maturing weapons program to successful construction, deployment, or detonation of a device."[6]

Institutional problems are an inevitable byproduct of state-building at a time of severe crisis. The fact that the power of the Kremlin is now limited and challenged from all sides is a major achievement of Russian democrats, but the chaotic and shaky conditions of new Russian governmental institutions has become a major long-term headache for arms control. Lines of domestic political conflict over arms control are tangled and shifting.

There is, first, the classic institutional conflict between executive and legislative bodies. The Russian State Duma, the lower house of the Russian parliament, has become a major vehicle of opposition to Yeltsin's arms control policies since it first convened in 1994. While its powers are constitutionally limited, it has repeatedly demonstrated its influence on the executive on issues of strategic, conventional, and chemical weapons.

Second, there is a complex jumble of interest groups within the executive branch, and a lack of a streamlined, effective policy-making mechanism on foreign policy issues, including arms control. The question, "Who makes policy?" is very difficult to answer with any precision on any given issue. Conflicting statements, shifting interpretations, unexpected initiatives, different agencies working at cross-purposes, and use of the media in bureaucratic infighting are everyday realities in the Russian Government. Sometimes, this conflict is consciously 'stoked up' by the executive to earn bargaining chips in international negotiations, but the main sources of intra-governmental struggles are objective and rooted in the crisis Russia is passing through. An especially dangerous aspect of the institutional weakness of the Russian state is inadequate civilian control over the military.

Political Trends

The Russian Federation is an emerging state which is only beginning to develop a sense of what its interests are, what place it occupies in the interna-

---

6. Estimates of the number of Soviet specialists in mass destruction weapons-related fields range from several hundred thousand to one million. R. Adam Moody, "Proliferation Implications of the Brain Drain," *Post Soviet Prospects* 4:12 (December 1996).

tional system, and what means it can and should use to protect and advance those interests. The new Russian elites, which led the fight for the dismantlement of the Soviet system, are alarmed at the continuing decline of Russia's power and influence. What they presented as "the triumph of democracy" increasingly looks like an historic defeat of Russia's statehood. Thus, the development of Russia's foreign policies is marked by a growing sense of insecurity, defensiveness, and suspicion.

The impact of the U.S. and other Western powers on the shape, outlook, and policies of the Russian Federation has been contradictory. Western governments have welcomed Russia's democratic and market reforms and given limited assistance to them, thus firmly associating themselves with both the successes and failures of reform policies. But Western help has become a matter of controversy in Russia — while the government is deeply disappointed at the paltry amounts of aid, opposition nationalists blame the West for the transition crisis, portraying "shock therapy" as a means of destroying Russia's economy and turning it into a Western raw materials appendage.[7] To make matters worse, Western policies have recently been evolving toward a kind of containment of Russia as a case of failing Westernisation and resurgent imperialism.

This puts Russian ruling elites in a bind. Since 1991, their formula for establishing control over post-Communist Russia has consisted of three main elements: building a new state, privatisation of the Soviet economy, and an alliance with the West. The ongoing deterioration of relations with the West, which may turn foreign policy from an asset into a liability, thus presents a direct and serious threat to the maintenance of the Yeltsin regime and compels it to look for other options.

The New Political Thinking, which defined Moscow's approach to international affairs since the late 1980s, is eroding and gradually giving way to other foreign policy orientations which may be summarized in the following points.

Western countries, especially the United States, have strong competitive interests vis-a-vis Russia. Despite the fact that Russia now has a capitalist-type economy, a pluralistic polity, and wants to cooperate with the West, the West

---

7. Polls taken by the Moscow-based Public Opinion Foundation in recent years have indicated that one Russian out of every three believes that Gorbachev's and Yeltsin's reforms are products of a Western conspiracy against Russia. *Interfax,* 7 February 1997. In a wide-ranging public opinion poll conducted in late 1995 by the Russian Independent Institute for Social and Political Studies, only 7 per cent of respondents agreed with the idea that "the West sincerely wants to help Russia," while 31 per cent endorsed the statement, "They want to weaken and turn us into a dependent state." However, 44 per cent shared a more pragmatic assessment of the West's goals: "They are helping us for their own gain, but it is quite natural." Rossijskij nezavisimyj institut sotsial'nykh I natsional'nykh problem. Massovoje soznanije rossijan v period obshchestvennoj transformatsii — real'nost' protiv mifov. Moskva, janvar' 1996, Part VI, p. 1. *National News Service,* 9 May 1997.

will cooperate with it only where both sides have common interests, and will compete where their interests clash. Elements of conflict are increasing in Russian-Western relations, and Russian leaders see Western powers trying to prevent the emergence of Russia as a strong competitor.[8] As an ideological underpinning of this perception, there is a resurgence of "Eurasianist" thinking which stresses the uniqueness of Russian civilisation and its distinctiveness from the West. While the present regime cannot afford to go for a break with the West, it is convinced that Russia can and should be more assertive in foreign policy while continuing its cooperation with Western countries.

Russia must realistically define its national interests, be guided by them, and be able to protect them. Debates on the definition of national interests can be expected to rage for quite a while before any stable system of views emerges, and this process is bound to be highly politicised and emotional. But there already exists a consensus among the elites that in the absence of an anchor like "national interest" Russian foreign policy is doomed to defeat. Another widely shared view is that it was a neglect of national interests that has led to the decline of Russian power in the past decade.

Military power is essential. If New Political Thinking postulated that the USSR overinvested in military power, the current Russian leadership is concerned about *too little* military power under Moscow's command. While the USSR spent 11–12 per cent of its GDP on defence in the last years of its existence, Russia now allocates about 5 per cent.[9] If one takes into account that Russia's GDP was roughly half that of the USSR and has declined by half since the start of Yeltsin's economic reforms, it turns out that in absolute terms, Moscow spends on defence about one-tenth of what it spent before the Soviet collapse. While the military and radical nationalists hold the most alarmist views on Russia's new military weakness, the view that Russia's defence potential is in crisis and must be rebuilt as soon as possible to increase Russia's clout in international affairs seems to be widely shared across the political spectrum.[10] In line with this perception, the Russian government

---

8. As Igor Ivanov, Russia's First Deputy Foreign Minister put it, "Influential political forces in some Western powers have not renounced their attempts to limit Russia's global role and reduce it to that of a regional power. This is manifested in their attempts to weaken Moscow's positions in such important areas as the Balkan Peninsula, the Middle East, and the Persian Gulf, as well as the former Soviet Union (especially in the Transcaucasian region and Central Asia)." *Nezavisimaya Gazeta,* 19 November 1996.

9. *Nezavisimaja gazeta,* 7 May 1997; Michael Gordon, "Yeltsin Dismisses His Two Top Military Leaders," *New York Times,* 23 May 1997.

10. In a classic statement on Russia's need for military power, State Duma Defence Committee Chairman Lev Rokhlin, a retired general, has written: "It goes without saying that mutually beneficial and friendly relations with neighbouring and more distant countries should be the basis of national defence and security. But history shows that partners respect the interests of only those countries which have a deterrence factor — a sufficient military potential. One example is the relations between

tried increasing defence expenditures in 1995, but revenue shortfalls in a depressed economy and IMF aid conditionalities have since forced further reductions in defence spending. According to President Yeltsin, it should be cut to 3–3.5 per cent of GDP.[11]

Nuclear weapons are more useful than before. As the country's conventional military strength declines and NATO moves closer to its borders, Russian leaders have rediscovered nuclear deterrence. In 1995, Russia officially abandoned the no-first use policy established under Gorbachev — now considered an unaffordable luxury. According to the new security doctrine approved by President Yeltsin in May 1997, Russia may use its nuclear weapons in response to an overwhelming conventional attack. Boris Berezovsky, Deputy Secretary of the policy-making Security Council, commented on the decision: "We are not speaking of making a first strike in order to secure advantage, but if we are driven into a corner and are left with no other option, we will resort to nuclear weapons."[12]

Russia must maintain its influence over the neighbouring post-Soviet states. If Gorbachev presided over the devolution of Moscow's power, and the early Yeltsin led the fight for the Soviet Union's dissolution, the dominant mood among Russian elites today is a sense that the so-called "near abroad," from the Baltic Sea to the Pamir Mountains, is an area of struggle for influence between Russia and the West, a struggle which Russia has been losing. The United States, in particular, is viewed as pursuing a policy of squeezing Russia out of its former rimlands, determined to get control over Caspian oil and other natural resources there and to keep Russia strategically contained. NATO enlargement is seen as a key element of this strategy. Increasingly, the main efforts of Russian diplomacy are focused on counteracting this perceived threat. The Russian-Belarus Union Treaty, signed in 1997, and Russia's renewed push for closer integration between member states of the CIS are among the manifestations of this policy.

---

the Soviet Union (with its giant military machine, the Warsaw Treaty Organization) and Western countries. Not a single important question was settled without consultations with the Soviet Union. But when Russia lost its military potential, the West started looking down on it and did not even deem it necessary to consult it before making the decision on NATO's eastward enlargement." *Nezavisimaya Gazeta*, 23 May, 1997.

11. *New York Times,* 23 May 1997. A recent Stockholm International Peace Research Institute study has noted a deeply contradictory approach of Russian elites to the role of military power: "...On the one hand, the financial burden of preserving a huge military potential is broadly recognized as being prohibitive for a country experiencing enormous economic hardships; on the other hand, a large military establishment is being presented as Russia's only claim to significant international status ..." *Russia and Europe: The Emerging Security Agenda*, Vladimir Baranovsky, ed. (Oxford: SIPRI, Oxford University Press, 1997), 7.

12. David Hoffman, "Yeltsin Approves Doctrine of Nuclear First Use if Attacked," *Washington Post,* 10 May 1997.

Russia must look for allies to counterbalance American hegemony. In a characteristic statement on the eve of his March 1997 meeting with U.S. President Clinton in Helsinki, President Yeltsin said, "Our diplomacy has made enough concessions. There is no more room to concede. Now it is up to the United States to yield, if we are to preserve partnership. I don't want a return to the Cold War, no, I don't want it. But then the world order must rest on the basis of equality. I am for a multipolar world. I am not in favor of the United States commanding everyone. Russia is a pole, the United States is a pole, then there are China and India, and there may be other poles, too."[13] Of course, in Helsinki, Russia did have to make a few more concessions in exchange for a few smaller ones on the part of the U.S., but a few weeks later, at a high-profile summit in Moscow, the presidents of Russia and China committed their countries to the creation of a new multipolar world order. In a classic balance-of-power approach, Russian elites are trying to enhance their country's international position by developing closer relations with those Asian countries which for one reason or another have had problems with the West. China, India, and Iran are seen as leading candidates for such a "counter-hegemonic" arrangement with Russia, and one of the consequences of this trend may involve strains on existing arms control regimes.

Clearly, these orientations represent an evolution from the Gorbachevian variety of liberal internationalism to a more conservative power-politics approach to world affairs — an approach which is increasingly perceived by Russian leaders as more realistic for the situation in which they find themselves at this historical juncture. It needs to be emphasized that what we have here are *trends* which the current Russian government would like to accomodate within the framework of continued Russian-Western partnership and without undermining existing arms control treaties. Yeltsin's 27 May 1997 agreement to accept NATO's enlargement plan after an unsuccessful Russian campaign to block it illustrates Moscow's reluctance to abandon its current Western orientation. Still, there are many signs that these trends are leaving their imprint on Russian foreign policy. The outcome of these changes will depend on a variety of factors, both internal and external. Western policies, whatever they are, are bound to play a major role in this process.

Arms Control and Prospects for Russia's Recovery

The Russian crisis presents a major challenge to the future of arms control. In the short-term perspective, maintenance of the existing arms control regimes requires a viable, effective state in Russia. In a medium- to long-term perspective, survival and development of the arms control system requires that Russia and the West continue and expand their cooperation in international security matters. It is a serious question whether it is possible to have both — a viable,

13. Interview prezidenta Rossii po povodu predstojashchej vstrechi s prezidentom SShA, *National News Service,* 17 March 1997.

effective Russian state and a continued Russian-Western security partnership. So far, the security partnership has coexisted with the Russian crisis, and it is not hard to imagine how a Russian recovery would be pursued on a path of "liberation" from Western influence.

The Russian crisis cannot continue forever. A search for workable recovery strategies is on, and the main debates revolve around the role and shape of the Russian state. In the Westernizing project undertaken since the mid-1980s, Russia has made a remarkable break with its centuries-old traditions of statehood, of which the USSR was only the latest variant — traditions of omnipotent bureaucratic power and a giant military-industrial complex maintaining an empire in the name of security. The new setup is a limited, chaotically pluralistic, market-oriented nation-state which seeks full integration with the world economy. Since this "atypical" Russian state is failing to provide for Russia's security, social peace, order, and prosperity, it is only natural to expect the traditions to reassert themselves in one form or another.

According to Russian communists and nationalists, Russia will never be able to recover while it stays bound to the West and its reform prescriptions, because the West supposedly wants to keep the Russians down, but also because the Western civilisation is rotten, bankrupt, and alien to the true Russian spirit. If these forces should prevail in the ongoing political struggles, all bets are off. A new dictatorship could emerge which would move to minimize Russia's dependence on Western countries, mobilise resources for a forced economic recovery, fight for the restoration of an empire, and challenge the West geopolitically and militarily. These political groups are openly hostile to arms control, as it is structured now. They claim that it is too expensive and reflects mainly Western priorities, while undermining important Russian national security interests. This perspective need not be adopted fully to become a significant influence on Russian policies, producing a long period of stalled negotiations and acrimonious compliance disputes.

In the meantime, there are corresponding trends on the Western side, as Cold War triumphalists — echoed by East European nationalists — urge the U.S. and its allies to take full advantage of Russia's predicament. A growing "Dump Russia" movement among American conservatives urges Western governments to worry less about Russian problems and concerns and actually dictate terms of cooperation to Russia — if the Russians do not accept them, let them stew in their own juices and stay outside of the Western world for the foreseeable future behind new NATO frontiers. This approach is based on the highly questionable assumptions that an economically weak Russia would not have any other policy options than continued cooperation with the West and demilitarisation, and that whatever contrary developments emerge in Russian politics can be safely checked by Western firmness.

Some of the basic ideas of arms control are now under attack in the United States. This development, symbolised by the fact that the ACDA now has to fight for its very existence as an institution, is rooted in a perception that as the single superpower the United States now has less need than before to coordinate its security policies with potential adversaries, especially when they are down. A fundamental tenet of New Political Thinking, the notion of mutual security, which postulated that a country, no matter how strong, could not hope to assure its security without negotiating with others in a way that enhances *their* security too, is challenged by latter-day American unilateralists. The seriousness of the challenge is reflected in a recent resolution of the U.S. Congress on ballistic missile defence. It gives a green light to the development of a national ballistic missile defence system which would 'explode' the ABM Treaty. The Administration's position that the conflict between BMD programmes and the Treaty must be negotiated with the Russians to obtain mutual agreement on the Treaty's modification, is rejected as superfluous.

It is easy to see how the two types of post-Cold War nationalism — one Russian, a product of new insecurity as a result of the loss of superpower status, the other American, a product of ill-conceived Cold War triumphalism and new arrogance of the single superpower — can interact and reinforce each other in a spiral which would spell great dangers for arms control and international peace. Meaningful and durable arms control is possible only if the parties have significant common perceptions of security. For the past ten years, unusually wide common ground existed between the USSR/Russia and Western powers, but it can no longer be taken for granted. There is a widening gap between Russian and Western perceptions — a gap which, if left unattended, can have very dangerous consequences. Vigorous efforts should be made by both sides to arrest this development. The East-West security consensus of the Great Decade should be rebuilt anew on the basis of new realities.

Negative interaction between Russia and the West can be countered. Russia's relapse into dictatorship and imperialism is not inevitable. It *is* possible to create a more active, more responsible, and more effective Russian state, which would help the Russian economy rebound, establish domestic order, and guarantee Russia's security *without* undermining the country's democratic prospects or creating threats for other countries. It is in the world's best interests that such a state does come into being, and the West has a crucial role to play.

(1) The West should be prepared to spend more to help demilitarise Russia. A strong case can be made for offering the Russian Government greater assistance for its efforts to fulfil its arms control obligations, deal with the disastrous situation in its armed forces, and implement effective military reform. The American CTR programme can be regarded as a model for a more comprehensive international effort, properly organised and supervised. The new NATO-Russia Council might well be the place where such a project could be developed.

Apart from its material value, this type of aid would help reassure Russia about Western intentions.

(2) Arms control itself, if it is perceived and pursued as an increasingly comprehensive and effective system for demilitarisation and stability on the global scale, should be regarded as vital for Russia's recovery from its crisis. Such an approach would call for a radicalisation of the arms control agenda. Russia will not remain in its militarily weakened condition if other powers develop stronger military potentials — so rather than wait for Russia to return to its military traditions, the West might engage Russia (and other states, too) in vigorous and far-reaching initiatives to reduce armaments, insitute confidence-building measures, and convert defence industries for peaceful use. The recent revival of international interest in denuclearisation and conventional disarmament, provided it gains official support, can be a helpful antidote for possible Russian remilitarisation.

(3) Some key tenets of arms control are in need of a review to account for new developments. For instance, there is growing support in both the United States and Russia for the creation of ballistic missile defences. Unless a new conceptual framework is devised jointly by the two countries to address the issues of offensive-defensive linkage, their unilateral actions are likely to derail strategic arms reductions. Another problematic area involves the shift from bloc-to-bloc arms control to more complex arrangements. If Russia finds itself confronted with an enlarged NATO acting as a bloc in arms control negotiations, they will be stalled. This situation calls for development of multilateral approaches and formulas.

(4) Will Russian arms control policies be based on voluntary choice or compulsion? Of course, the choice is never entirely voluntary for any country. There is an important difference, however, between Moscow making concessions to the other superpower, deemed to be roughly equal in power for the sake of a reasonable trade-off, and Moscow being forced to accept arrangements it openly objects to because its reduced power status does not entitle it to better terms. The temptation to keep Russian power as limited as possible is quite strong, especially among East Europeans, who have an historical fear of Russian imperialism. But this temptation is dangerous, since yielding to it evokes impulses in the Russian body politic exactly opposite to what is desirable. The USSR's exit from the Cold War, liberation of Eastern Europe, and the demise of the USSR resulted from voluntary choices made by Moscow elites. This is hardly a just case for imposing a punitive peace.

(5) In a wider political context, if the West wants Russia to follow a non-threatening recovery path, it should avoid policies which increase the

Russian sense of insecurity. That refers, first and foremost, to NATO's policies in Eastern Europe and new independent states of the former USSR. NATO enlargement is an exercise in brinkmanship. Now that Russia has conditionally accepted it, unusual prudence is required of the West to make sure that this process is not detrimental to Russia's security. Unless NATO undergoes significant transformation away from its original Cold War mode, it is likely to be perceived as hostile to Russia. Extending NATO membership to Baltic states and Ukraine would be a particularly dangerous step.

The West has a tremendous preponderance of power *vis-a-vis* Russia in crisis. This gives Western leaders a great deal of room to develop responsible policies, but also creates a sense of complacency in the West which weakens the incentive to critique the status quo and seek new solutions to problems. Dealing with the Russian crisis is the toughest test of Western statesmanship today.

Chapter Eighteen

Future Challenges for Multilateral Arms Control: A Case Study on Korea

George Lindsey and Jim Bayer*

Background

This chapter outlines two related studies which are being prepared for the sixth annual Canada-Korea North Pacific Arms Control Workshop. The workshops, which are jointly sponsored by the Verification Research Program at the Canadian Department of Foreign Affairs and International Trade and the Korean Institute for Defence Analysis, are held alternately in Victoria B.C. and Seoul. The 1997 workshop will be held in Seoul in May. Over the years, the workshop has expanded from being a bilateral affair to include participants from China, Japan, Russia, the United States, as well as Canada and South Korea. The theme of the workshop will be "promoting peace and security on the Korean Peninsula: preventing regional conflicts." As part of Canada's preparation for participation in this workshop, two research papers are being undertaken: one by Dr. J. Bayer summarises the perspectives of South and North Korea on promoting peace on the Korean Peninsula through arms control. The second, by Dr. George Lindsey, summarises the perspectives on Korean arms control through the eyes of extra-Peninsula players.

The Process

Both papers are structured around a number of key questions and issues. These include, but are not limited to:

- determining what each player sees as the desirable state of affairs on the Korean Peninsula in the short, medium, and long terms;
- describing those issues which each side feels must be resolved in order to achieve the desired state of affairs;
- examining how each side addresses conventional arms control and weapons of mass destruction;

---

* This chapter is an edited version of a presentation made to the Fourteenth Annual Ottawa NACD Verification Symposium, "Cyberspace and Outer Space: Transitional Challenges for Multilateral Verification in the 21st Century," 12–15 March 1997, Montebello, Quebec, Canada.

- determining each party's attitude toward verification;
- assessing the extent to which political conciliation on the Peninsula can be achieved; and,
- assessing the role that can be played by other countries in the region to help achieve inter-Korean reconciliation.

To facilitate the analysis of these key issues and questions, a two-step process was adopted. First, in order to develop more "authentic" perspectives on the arms control issues involved in the analysis, regional experts were commissioned to write short "perspective" papers outlining the attitude of their particular country toward the issues associated with promoting peace and security on the Korean Peninsula through arms control. For example, for the Peninsula paper, regional experts from North and South Korea were to provide short papers outlining the perceived view of their governments, while regional experts from China, Russia, Japan, and Canada were to provided papers outlining the perspectives of the non-Peninsula players. Initially, in the case of North Korea, a number of "proxies" were also asked to analyse Pyongyang's position on Peninsular arms control. This was done because there was some concern that it might prove difficult to find a North Korean national who could provide such an analysis. In the end, however, such a person was identified. There is no insistence that the views of one regional expert be in agreement with the views of the others. Each contributor was responsible for her/his own paper.

The second stage in the process will be for the two principal authors to take the perspective papers and synthesise them into separate research papers: one reflecting the views of the Peninsula players, the other the views of non-Peninsula players. In carrying out this task, the major objective of each author is to go beyond summarising the positions contained in the regional experts' papers. The wider purpose of these "overarching" papers will be to identify areas of agreement and disagreement among the states involved, and seek to establish a common basis for discussion between them.

Preliminary Findings Regarding the Koreas' Perspectives

A number of observations will be made with regard to the North Korean position on the issue as well as on the complex relationships involving Peninsular and non-Peninsular players.

North Korean Position

The following synopsis of the North Korean position on the prospect for promoting peace and security on the Korean Peninsula through arms control is based on the paper received from the North Korean regional expert living in Japan. It is not based on the positions outlined by North Korean "proxies." It is interesting in its own right for this reason.

Desirable State of Affairs on the Korean Peninsula: The desirable state of affairs on the Korean Peninsula was implied to be political unification between the North and the South.

Issues Which Need Resolution: The issues which require resolution before the desired state of affairs could be achieved were identified as follows:

- the signing of a peace treaty between the United States and North Korea to replace the 1953 Armistice;
- the normalisation of U.S.-North Korean relations; and,
- the formation of a "coalition" government in South Korea that includes political elements not hostile to cooperation with North Korea.

Arms Control Measures: Once the issues related to a peace treaty, the normalisation of relations, and the formation of a coalition government have been resolved, a number of arms control measures could be implemented by the South-North Joint Military Committee and Joint Military Commission, as established under the terms of the 1992 North-South Basic Agreement.

Essentially, the arms control measures proposed fall into two categories:

(i) *Immediate measures*

First, there were those measures which it was felt needed to be immediate once the preconditions for arms control negotiations had been met. Generally, these were related to the Demilitarized Zone (DMZ) and involved arms reduction, as opposed to transparency and confidence-building measures such as exchange of military data, personnel, or observers. The proposals included:

- detargeting of all North and South Korean missiles aimed at each other;
- removing artillery and rockets from the border area;
- dismantling all military equipment and facilities in and along the DMZ;
- removing all mines from the DMZ and adjacent areas; and,
- ending all military manoeuvres.

(ii) *Longer term measures*

The longer term arms control proposal envisaged a four-stage, ten year plan to reduce troop levels on both sides of the frontier to 100,000 each.

- Year I: both sides make preparations for reductions.
- Year II–IV: both sides reduce forces by one third.
- Year V–VII: both sides reduce forces by a further one third.
- Year VIII–X: both sides reduce forces by a final one third to 100,000.

Weapons of Mass Destruction: The North Korean position was that this issue should not be a problem since neither side possessed weapons of mass destruction, although the option for North Korea to build nuclear weapons remained open in the event that the United States did not live up to the terms of the October 1994 framework agreement.

Verification: The North Korean proposal recognised the need to employ a variety of verification techniques in order to implement such an agreement. The specific verification proposals reflect North Korea's apparent willingness to accept on-site inspection by monitors from South Korea and international agencies, as well as the use of third-party intelligence. The latter point was surprising, given the North Koreans' response to the IAEA's use of third-party intelligence on the nuclear question in 1993.

The specific proposals include: on-site inspection by joint teams of observers to verify the dismantling of military bases; UN observers to monitor troop reductions; international observers to monitor reduction of military equipment; U.S., Russian, and French "spy satellites" to monitor implementation of the agreement.

Role Played by Non-Peninsular Powers: The North Korean proposal saw the United States as the principal non-Peninsular power having a role to play in bringing a resolution to the Korean issue. This role reflected Washington's historical support for the South Korean government, the presence of 37,000 U.S. troops on the Korean Peninsula, and its superpower status in the region. Japan was seen as having a role in helping to broker peace between the United States and North Korea once Tokyo apologised and paid compensation to North Korea related to its past 'misdeeds' and atrocities. Interestingly, China and Russia were seen as having no role to play other than to support the goal of reconciliation between North and South Korea.

Chapter Nineteen

The Multilateral Dimensions of the 'Korean Problem'

George Lindsey*

Introduction

Most of the negotiations which led to the earlier important breakthroughs in arms control, such as the SALT, INF, and START treaties, were bilateral, carried out between the United States and the Soviet Union, and concentrated on armaments in Europe and North America. Meanwhile, some of the less powerful countries, such as Canada, have devoted their efforts on arms control toward multilateral approaches. While the verification of bilateral treaties is a matter for the two signatories, every member of a multilateral agreement is concerned with its verification.

The threats to security in Northeast Asia posed by the situation on the Korean Peninsula, offer unusually complicated problems for arms control, particularly pressing requirements for the verification of any measures which may be agreed, and a need for confidence-building when the confidence is justified. Although the principal actors are two in number — the ROK and the DPRK (respectively referred to as South and North Korea in the following text) — any comprehensive analysis of the situation and the problems to be faced must be multinational.

There have been efforts to use game theory in the analysis of problems such as this. Most of the classical fundamentals of game theory, and the majority of its more useful methods and instructive conclusions, have been based on two-person zero-sum games with perfect information. But the 'Korean problem' is better modelled by a game in which the success of arms control would produce an outcome which would have a positive value for each of many players, rather than zero-sum for two players. And unless effective verification were established, it would be a game of imperfect information.

---

* This chapter is an edited version of a presentation made to the Fourteenth Annual Ottawa NACD Verification Symposium, "Cyberspace and Outer Space: Transitional Challenges for Multilateral Verification in the 21st Century," 12–15 March 1997, Montebello, Quebec, Canada.

The External Players in the Game: Their Recent History Relating to Korea

The central players are the two Koreas, but both have made longstanding declarations of intent to unify into a single state. There are profound differences as to the terms under which unification would be acceptable, and doubts as to the real desire on the part of either of the two principals or of their neighbours for unification to be accomplished in the immediate future.[1] For analysis of the prospects for unification and for arms control there are two central players, but in identifying the general security interests of both these two and the other players there must be consideration of the situation that would exist without unification, or after unification, should it occur.

The three powerful neighbours of Korea are China, Japan, and Russia. North Korea shares a long land frontier with China, and all of Korea was part of the Manchu Empire until 1895. North Korea has a tiny common border with Russia, close to the vital Russian zone around Vladivostok, and all of Korea came under Russian protection in 1897, after its separation from China. After Japan's defeat of Russia in 1905, Korea became a Japanese protectorate, and was annexed by Japan in 1910, only achieving independence 35 years later, at the end of the Second World War.

The major external power in the game is the United States. Although located far away from Asia, it has waged three hot wars and one cold war on the northwestern coast of the Pacific Ocean in the last half century. The hot wars were against Japan, North Korea backed by China, and North Vietnam. Russia was an ally against Germany in both World Wars, but the principal opponent — in both Europe and Asia — in the Cold War. While by no means a major player, Canada has had the same alignments as the U.S. in the World Wars, the Korean War, and the Cold War, although it did not participate in Vietnam. Another player with a peripheral role in the 'Korean problem' is Taiwan, mainly through its importance for the relations between China and the United States.

Pairs of Countries with Predominately Good Relations

A difference between the security strategies of Asian and Western states is their willingness to enter into alliances. Asian states have usually chosen to stand alone. However, among the eight players being considered here, the U.S. has formed alliances with Japan, South Korea, Taiwan, and Canada. Less formal, but nevertheless significant, Canada has excellent relations with South Korea — with which it was a co-belligerent in the Korean War — and with Japan. While not as close as they were during and following the Korean War, the relations between North Korea and China are generally cooperative.[2]

---

1. Nicholas Eberstadt, "Hastening Korean Reunification," *Foreign Affairs* 76:2 (March/April 1997), 77–92.

Thus, of the 28 pairs among eight players, there are seven on distinctly and persistently friendly terms. Or, leaving out Canada and Taiwan and counting only the six central players, three out of the fifteen pairs are on consistently good terms. This is not a situation in which disarmament, arms control, or even confidence-building is going to be easily introduced.

Pairs of Countries with Predominantly Bad Relations

North Korea has set itself on a path of determined self-sufficiency and isolation. Its deteriorating economy has motivated it to accept some assistance with food supplies, and some channels of discussion have been opened up in recent years. But, on balance, it seems that the relations between North Korea and South Korea, Japan,[3] and the United States,[4] should be characterised as bad. In all three cases, there is a classic conflict of political ideology. The legacy of Japanese exploitation of Korea and Japanese concerns over North Korea's efforts to acquire missiles able to deliver weapons of mass destruction to Japan are factors — offset to some degree by the large number of North Korean citizens employed in Japan and sending much needed hard currency back home. Hostility toward the United States is due mainly to the American support of South Korea, both during the Korean War and with the continuing stationing of U.S. troops in South Korea. Somewhat detached from the Korean problems is the hostility of China towards Taiwan, as an aftermath of the Chinese revolution, of the continuing deep seated quarrel over the separation of Taiwan from mainland China, and of ideological differences.

Thus, out of the 28 pairs of states, there are four with consistently bad relations. Or, of the six central states, three — each including North Korea — out of fifteen pairs are persistently unfriendly. For them to place any confidence in arms control agreements there would have to be highly reliable verification in place.

Pairs of Countries with Mixed Relations

If we confine the enumeration of mixed relations to the six central players, and subtract the three good and three bad pairs already identified, this leaves nine out of the fifteen which can be described as mixed.

North Korea and Russia

North Korea's relations with Russia are far less cordial than was the case when Russia expelled Japan from North Korea in 1945, and then departed in

---

2. Robert A. Scalapino, "The Major Powers and the Korea Peninsula," *Korean Journal of National Unification* 3 (1994), 9–47.
3. Paul Bracken, "Risks and Promises in the Two Koreas," *Orbis* 39:1 (Winter 1995), 55–64.
4. Manwoo Lee, "North Korea: The Cold War Continues," *Current History* (December 1996), 438–442.

1947, during the Korean War, or the longer period after that during which Russia was supplying arms on favourable terms. The North Korean leadership observes with apprehension the collapse of the USSR — the central bastion of Communist ideology — and the poor state of the Russian economy. Russia would not want another nuclear power so close to its border. But neither would it want to see North Korea collapse and be absorbed into a unified single Western-oriented Korean state.[5]

South Korea

While North Korea's relations with Russia may be cooling, those of South Korea are improving.[6] Trade between the two is growing.

Whereas South Korea had every reason to consider China as a major enemy during and for many years after the Korean War, relations have been established and mutually profitable trade is increasing.[7] China probably sees a solution to the 'Korean problem' as a means of preventing the appearance of a new nuclear weapons state on its border.

South Korea shares with the North resentment of former treatment by Japan, and has a disagreement regarding sovereignty over some small islands — Takeshima Rocks in the Sea of Japan and the Danjo Islands in the East China Sea. Nevertheless, both South Korea and Japan share concerns over North Korean armament, and wish to retain the presence and backing of the United States in Northeast Asia.[8]

China

China's relations with Russia, with which it shares a long and frequently contested land border, have undergone a number of drastic changes. During the nineteenth century Russia seized control of several pieces of territory which China still claims today to have been unfairly appropriated in "unequal

5. "North Korea: A Country in Transition," *International Security Review* 1996 (London: Royal United Services Institution for Defence Studies, 1996), 357–383.

6. Alexander I. Nikitin, "From Red Star to Two-Headed Eagle: Translating Cross-Recognition from Soviet into Russian," *Korean Journal of National Unification* 3 (1994), 79–95.

7. James T. Myers, "Issues and Prospects for Cross-Recognition: A Chinese Perspective," *Korean Journal of National Unification* 3 (1994), 97–110.

8. Tetsuro Doshita, "Japan's Involvement in Meeting Proliferation Concerns in the North Pacific," in *Confidence Building in the North Pacific: New Approaches to the Korean Peninsula in the Multilateral Context. Proceedings of the Canada/Korea North Pacific Arms Control Workshop 1996* (Victoria: University of Victoria and Korean Institute for Defense Analyses, 1996), 187–196. See also Byung-joon Ahn, "Regionalism and the US-Korea-Japan Partnership in the Asia-Pacific," in *US-Korean Relations at a Time of Change*, Report of a conference held in Seoul in September 1994, organized by the Center for Strategic and International Studies, Washington, and the Research Institute for National Unification, Seoul, 99–130.

treaties." At the beginning of this century Russia occupied Manchuria, but was driven out by Japan in 1905. In 1931 Japan invaded Manchuria and subsequently established the state of Manchukuo. During the last days of the Second World War, Russia entered Manchukuo and defeated the Japanese. However, after the war was over, Russia left the area to be returned to China. Russia did not support the Chinese Revolution, but after the establishment of the People's Republic of China in 1949 Russia aided in the industrial development of China until 1960, at which time an ideological conflict resulted in a great estrangement and a sudden abandonment of many major uncompleted projects in China. Fighting broke out in 1969 over the precise location of several sections of the border. Relations gradually emerged from this nadir in the 1980s.[9]

Today, both China and Russia face the possibility of internal political upheavals. However, China's economy is growing fast and the country has a need for stability. The border disputes between the two abated in 1991. Both have reason to desire stability and arms reductions in Korea.

Of all the multitude of historical reasons for resentment between states in Asia, China has the longest list. There is plenty to resent regarding Western exploitation, but no less for the behaviour of Japan. However, while these indignities may not be forgotten, there is good cooperation in investment and trade. China and Japan have no difficulties over land borders, but the East China Sea holds the potential for disputes regarding sovereignty over small islands, such as the Senkakus, which could involve rights to fishing and to possible oil reserves. Since the Korean Peninsula is practically surrounded by China and Japan, these two countries hold dominating geostrategic power over activities on the Peninsula.[10]

China's relations with the United States cover an immense agenda, of vital significance to the whole world and extending far beyond the security problems of Korea. As China grows in prosperity and power, it is natural for the United States to foresee an Asia in which China will be playing a role far more influential than has been the case in centuries past. One can detect in the United States significant support for two sharply contrasting approaches to a future China policy. One is often labelled as "engagement," extending cooperation on many fronts, such as seeking liberalisation of trade restrictions, encouraging the participation of China in international fora, and discouraging its arms exports. The opposite approach is "containment," regarding China as a growing rival for influence and business, a competitor for strategic power, and a delinquent in need of reform in the areas of human rights and respect for copyrights.[11]

---

9. Gilbert Rozman, "A Regional Approach to Northeast Asia," *Orbis* 39:1 (Winter 1995), 65–80. See also William E. Odom, *Trial After Triumph: East Asia After the Cold War* (Hudson Institute, 1992).
10. See Doshita, *op cit.*, and Rozman, *op cit.*
11. See Richard Bernstein and Ross H. Munro, *The Coming Conflict with China* (New

A serious bone of contention exists over the status of Taiwan, which, while clearly unable to achieve its goal of making a victorious return to the mainland, demands international recognition as an independent state. Beijing is adamantly committed to reunification into "One China," and rejects relations with countries which recognise Taipei. Mainland China bombarded the Taiwanese Islands of Matsu and Quemoy in the 1950s, and as recently as 1996, conducted aggressive demonstrations of missile testing close to Taiwan.[12] The United States is friendly toward Taiwan, provides it with modern armaments, and, on occasion, demonstrates its support with the presence of powerful naval forces.

However, as concerns the problems of Korea, China has changed from being the ally of North Korea against South Korea and the U.S., and is working to prevent the arming of North Korea with weapons of mass destruction. Although China may hope for an eventual American withdrawal from Asia, it probably favours the continued presence of U.S. troops in South Korea and Japan as a stabilising factor for the immediate future.

Russia

The relations between Russia and the United States, adversarial throughout the Cold War, have improved to a remarkable degree. Serious Russian internal problems, political as well as economic, absorb most of the attention of the leadership, which seeks stability on the turbulent Russian frontiers. Like China, Russia has changed from being a backer of North Korea against the South to a neighbour which wants peace in Northeast Asia.

The relations between Russia and Japan have generally been antagonistic, since the Russo-Japanese War when Japan gained the southern half of Sakhalin Island, which has been a bone of contention ever since. In 1941, at a time when Russia was concerned over its eastern front and Japan was planning expansion in Asia, a non-aggression pact was signed, which was observed until the last days of the Second World War, in spite of the fact that Russia was fighting on the side of the Western Allies in Europe and Japan against them in Asia. In the last week of the war Russia invaded Manchuria and Korea, and occupied Sakhalin and the Kurile Islands.

After the end of the Second World War, Japan made a miraculous transition into the Western camp while Russia departed. Japan served as a base for the UN operations in the Korean War, while Russia supported North Korea.

The collapse of the Soviet Union has eased this animosity, and there has been modest investment of Japanese capital and expertise in the Russian Federation. Both countries want a stable Korea, without weapons of mass

York: Alfred Knopf, 1997), and David Shambaugh, "Containment or Engagement of China?" *International Security* 21:2 (Fall 1996), 180–209.

12. See Bernstein and Munro, *op cit.*

destruction. Japan still seeks return of four of the Kurile Islands and the southern part of Sakhalin Island, and there are disputes regarding fishing rights.[13]

Although Russia has a long eastern coastline, its usable seaports are separated from the open Pacific Ocean by the Sea of Japan and the Sea of Okhotsk, or, farther north, face waters bound by winter ice. The exits to the Pacific lead between South Korea and Japan — through the Korea Strait — between Japanese Islands, between Japan and southern Sakhalin Island, or between islands of the Kurile chain. It is easy to see why Russia is sensitive to the control of these "choke points."

Conclusion

The 'Korean problem' can be modelled as a 2+4 person non-zero-sum game with imperfect information. Among the six players, there are some coalitions for specific purposes, but each has different bilateral considerations with each of the others. The history of the region does not provide a promising base on which to build confidence, and it seems certain that none of the players will be attracted to disarmament or arms control unless very reliable measures are put in place for the verification of compliance with any agreements that may be concluded.

13. Masahiko Asada, "Japan and a New Drift toward Confidence Building in the Asia Pacific Region: Retrospect and Prospect," in *Confidence Building in the North Pacific: New Approaches to the Korean Peninsula in the Multilateral Context. Proceedings of the Canada/Korea North Pacific Arms Control Workshop 1996* (Victoria: University of Victoria and Korean Institute for Defense Analyses, 1996), 121–138. See also Scalapino, *op cit.*, and Rozman, *op cit.*

Chapter Twenty

Symposium Summary

Jacqueline Simon*

Participants in the 14th Annual Ottawa NACD Verification Symposium used the conference as an opportunity to look into the next millennium and speculate about what it may hold for arms control and disarmament. The topic of the conference sparked a great deal of innovative thought and those presenting papers revealed an exciting and challenging future for arms control and disarmament in the 21st century. Despite a seemingly diverse range of topics, a number of common threads could be traced through the presentations. These overarching themes will be explored in this brief summary of the conference.

One of the issues which emerged on numerous occasions was that of universality and the constant tension between the 'haves' and 'have nots' of the nuclear world. This issue was raised in a number of different contexts. First, it was asserted that an effort needs to be made by the recognised nuclear possessors to live up to their NPT obligations. This is necessary to ensure that aspiring proliferators and *de facto* possessors will also live up to their NPT obligations. The assertion was made during the conference that universality costs very little to promote but could have great benefits. It was suggested that perhaps a global treaty on missiles would be sufficient to reassure the nuclear possessors so that they could begin to meet their NPT obligations.

There was speculation that the gap between the 'haves' and the 'have nots' will continue to grow as new information technology is introduced. This gap means that when the 'haves' are ready to control a technology, the 'have nots' view that activity as discriminatory as only they have just acquired the technology. For them, non-proliferation often looks like containment. This ties back to the necessity to promote universality and to illustrate that there is not a bias against arms control among the nuclear possessor states.

Another issue raised repeatedly was the feeling that existing treaties are insufficient to deal with the changing proliferation threat and that these treaties need to be either revised or supplanted by new treaties. The feeling was

---

* This chapter is an edited version of rapporteur notes recorded and compiled by Jacqueline Simon at the Fourteenth Annual Ottawa NACD Verification Symposium, "Cyberspace and Outer Space: Transitional Challenges for Multilateral Verification in the 21st Century," 12–15 March 1997, Montebello, Quebec, Canada.

also expressed that some of the terminology which frames the discourse on arms control and disarmament is becoming antiquated and should be reviewed and revised in order to stay current and meet the new proliferation threat. The existing non-proliferation and arms control regimes were created during the Cold War to combat a threat which no longer exists. The greatest danger of proliferation today is not vertical but horizontal. New regimes and treaties must take account of the non-state actors which play an increasing role in the proliferation and non-proliferation processes.

The increasing role of non-state actors has important ramifications for the way verification and arms control are conducted. The distinction between national and international security is becoming increasingly blurred in a world in which threats may be subnational and transnational, as well as national. This changed nature of the threat necessitates a more multilateral approach to non-proliferation, arms control, and verification. Participants at the conference emphasised the need for common solutions to common problems. It was asserted that there must be more constructive dialogue between nuclear possessors and those without nuclear weapons in order to achieve multilateral efforts to address these problems.

One of the areas where participants predicted increasing future cooperation was in verification activities. Verification will likely be revolutionised in the 21st century. As states strive to make their activities more cost-effective, this will induce greater cooperation and exchange of information for the purposes of verification. In addition to cooperation between states, non-state commercial enterprises will become increasingly involved in such verification activities as satellite imaging, as they can perform these activities more cost-effectively. The involvement of commercial enterprises means that this information will be more readily accessible to parties previously without access.

Verification will also be changed in other ways. The information revolution and the technology driving it present both opportunities and challenges for verification. Increased knowledge and information about proliferation activities leads to a new awareness of compliance problems which must be addressed. In addition, the technology used to prevent the spread of weapons may also be used in their proliferation. The links between states provided by information technology may provide new security vulnerabilities in addition to their positive aspects.

Several participants emphasised that, although the technical difficulties of disarmament and verification may be reduced by new technology, proliferation activity does not occur in a political vacuum. Both proliferation and the efforts made to combat it are largely determined by the political will driving each activity. Although the proliferation threat of the future may transcend boundaries, the solutions to this problem often run into state-based political problems. Politics seems unable to keep pace with rapid technological change as we move into an age unbound by space or time.

Yet, in order to resolve the proliferation problem, political aspects of the situation must be addressed. Advances on the technical side of non-proliferation are insufficient to resolve the problem alone. One way in which these political problems may be avoided is via the increasing involvement of non-state actors working multilaterally to address the problem. At the conference, the landmines case was presented as a successful example of looking beyond state-based solutions to a problem. In this case, non-state actors used the internet and other media available to them to spread public awareness of the problem and to promote an international solution.

On the subject of outer space, there was a strong consensus among participants that it is essential to ensure that it remains weapons free. If it does not, a vast new arena for proliferation activities will be opened up and international security will be immeasurably threatened. Participants stressed the importance of the Outer Space Treaty and asserted that the exploration and use of outer space should be to the benefit of all humanity. There is a great deal of positive potential in outer space in terms of increased communications and verification and it is these activities which should be encouraged.

It appears that the next century holds a number of challenges and possibilities for NACD verification. The information revolution, with its rapid technological advancement, will provide avenues for both proliferation and its prevention. New and old proliferation threats will co-exist in the 21st century. Similarly, both new and old verification measures will be necessary and will be complementary when used in tandem. In order to meet the proliferation threat of the next century, new technology must be adopted and old treaties and regimes must be revised in order to keep them current and applicable to the threats we face in a rapidly changing world.

List Of Abbreviations and Acronyms

ABM	Anti-Ballistic Missile Treaty
ACDA	United States Arms Control and Disarmament Agency
ASEAN	Association of Southeast Asian Nations
BMVC	Baghdad Monitoring and Verification Center
BTWC	Bacteriological (Biological) and Toxin Weapons Convention
CBM	confidence-building measure(s)
CCW	Convention on Certain Conventional Weapons
CD	Conference on Disarmament
CFE	Treaty on Conventional Armed Forces in Europe
CMC	Cooperative Monitoring Center
CSBM	confidence- and security-building measure(s)
CTBT	comprehensive test ban treaty
CWC	Chemical Weapons Convention
DPRK	Democratic People's Republic of Korea
Fizz	fissionable material "cutoff" treaty
GII	global information infrastructure
GPR	ground penetrating radar
GPS	Global Positioning System
GSE	Group of Scientific Experts (CD)
HEU	highly-enriched uranium
HUMINT	human intelligence
IAEA	International Atomic Energy Agency
IMS	International Monitoring System
INF	Intermediate-Range Nuclear Forces Treaty
ISMS	International Seismic Monitoring System
ISR	intelligence collection, surveillance, and reconnaissance
LTBT	Limited Test Ban Treaty
MFO	Multinational Force Observers
MIRV	multiple independently targeted re-entry vehicle
MTCR	Missile Technology Control Regime
NACD	non-proliferation, arms control, and disarmament
NAM	Non-Aligned Movement
NATO	North Atlantic Treaty Organization
NBC	nuclear, biological, chemical
NGO	non-governmental organization
NNWS	non-nuclear weapons states
NORAD	North American Aerospace Defence Command
NPT	Nuclear Non-Proliferation Treaty
NSG	Nuclear Suppliers' Group
NTM	national technical means

NTSC	non-traditional security challenges
NWS	nuclear weapons states
NWFZ	nuclear-weapons-free zone
OAS	Organization of American States
OHI	overhead imagery
OMV	ongoing monitoring and verification
OPCW	Organization for the Prohibition of Chemical Weapons
OSCE	Organization for Security and Cooperation in Europe
OSI	on-site inspection(s)
P-5	Group of 5 Permanent Members of the UN Security Council
PCS	personal commications systems
REVCON	Review Conference
SCADA	supervisory control and data acquisition
SFM	Sinai Field Mission
START	Strategic Arms Reduction Treaty
UNDC	United Nations Disarmament Commission
UNEF II	United Nations Emergency Force
UNSCOM	United Nations Special Commission on Iraq
UNSSOD IV	United Nations Fourth Special Session on Disarmament
VEREX	Ad Hoc Group of Governmental Experts, BTWC
VICS	Verification and Implementation Coordination Staff, NATO
WMD	weapons of mass destruction

Other Publications by the Centre for International and Security Studies

The following is a listing of monographs published by the Centre for International and Security Studies dealing with non-proliferation, arms control, disarmament, and verification issues. For information on how to obtain these publications, please contact: Centre for International and Security Studies, York University, 4700 Keele Street, North York, Ontario, Canada M3J 1P3, Tel: (416) 736-5156, Fax: (416) 736-5752, E-mail: yciss@yorku.ca, Internet: http://www.yorku.ca/research/ciss

Non-Proliferation Agreements, Arrangements and Responses. Proceedings of the 1996 Canadian Non-Proliferation Workshop, Andrew Latham, ed., 1997.

Verification, Compliance and Confidence-Building: The Global and Regional Interface. Proceedings of the Thirteenth Annual Ottawa NACD Verification Symposium, J. Marshall Beier and Steven Mataija, eds., 1996.

Multilateral Approaches to Non-Proliferation. Proceedings of the 4th Canadian Non- Proliferation Workshop, Andrew Latham, ed., 1996.

Proliferation in All Its Aspects Post-1995: The Verification Challenge and Response, J. Marshall Beier and Steven Mataija, eds., 1995.

Moving Beyond Supplier Controls in a Mature Technology Environment. Proceedings of the Third Canadian Non-Proliferation Workshop, David Mutimer, ed., 1995.

Non-Proliferation and Multilateral Verification: The Comprehensive Nuclear Test Ban Treaty (CTBT), Steven Mataija, ed., 1994.

Reconsidering Confidence and Security Building Measures: A Critical Analysis, Andrew Richter, 1994.

Control But Verify: Verification and the New Non-Proliferation Agenda, David Mutimer, ed., 1994.

Proliferation and International Security: Converging Roles of Verification, Confidence Building and Peacekeeping, Steven Mataija and Lyne C. Bourque, eds., 1993.

Multilateral Verification and the Post-Gulf Environment: Learning From the UNSCOM Experience, Steven Mataija and J. Marshall Beier, eds., 1992.

Towards Helsinki 1992: Arms Control in Europe and the Verification Process, Heather Chestnutt and Steven Mataija, eds., 1991.

International Verification Organizations, Ellis Morris, ed., 1991.

Verifying Conventional Force Reductions in Europe: CFE I and Beyond, Heather Chestnutt and Michael Slack, eds., 1990.